D1300192

Second International Mathematics Study Reports Now Available

(see order information below):

Summary Report: Provides an overview of major U.S. findings of the Study, with 41 graphs and many tables. 130 pp.

Detailed Report: A comprehensive compendium of findings on the teaching and learning of mathematics in U.S. schools at the eighth and twelfth grade levels. 440 pp.

The Underachieving Curriculum: *Assessing U.S. School Mathematics from an International Point of View:* A non-technical presentation of the major findings of the Study, with a discussion of their implications for the improvement of U.S. school mathematics. Contains over fifty figures, tables, and accompanying commentary (see sample excerpts in this *Forum* feature). 120 pp.

Technical Report I: A technical presentation of U.S. cognitive data, including item-level student response patterns, both pretest and posttest, and international scores. 270 pp.

ORDER FORM

PLEASE PRINT OR TYPE

To: Stipes Publishing Company
    10-12 Chester Street
    Champaign, IL 61820
    Tel: (217) 356-8391

Please send me the following report(s) on the Second International Mathematics Study:

From: (Name) _____

     (Street Address) _____

     (City) _____ (State) _____ (Zip) _____

     Tel: ( ) _____

	Qty.	Unit Cost	Total
United States Summary Report (130 pp.)		$7.80	
United States Detailed Report (440 pp.)		$21.00	
The Underachieving Curriculum (120 pp.)		$8.00	
Technical Report I (270 pp.)		$9.00	
Subtotal			
Postage & Handling			$2.00
Total amount enclosed			

Orders of 10 or more copies (titles may be combined) are subject to a 15% discount.

THE UNDERACHIEVING CURRICULUM:

ASSESSING U.S. SCHOOL MATHEMATICS
FROM AN INTERNATIONAL PERSPECTIVE

Curtis C. McKnight, University of Oklahoma

F. Joe Crosswhite, Northern Arizona University

John A. Dossey, Illinois State University

Edward Kifer, University of Kentucky

Jane O. Swafford, Northern Michigan University

Kenneth J. Travers, University of Illinois at Urbana-Champaign

Thomas J. Cooney, University of Georgia

STIPES PUBLISHING COMPANY
10 - 12 Chester Street
Champaign, Illinois 61820

The U.S. National Coordinating Center for the Second International Mathematics Study was located in the Department of Secondary Education, College of Education, University of Illinois at Urbana-Champaign.

— Kenneth J. Travers, Director

Center staff and National Committee members are listed in Appendix I. University of Illinois personnel who assisted in the production of this report include:

Copy Editing
Scott F. Stoddart, Department of English

Technical Assistance
Chantanee Indrasuta, College of Education

Word Processing
Del Jervis, College of Education

Typesetting
Judi Kutzko, Stenographic Services

Cover Design and Graphics
Lisa Bechtel, Instructional and Management Services

Art/Paste-up (Stipes Publishing Company)
Laura Jolley

Ninth Printing, April 1990
ISBN 0-87563-298-X

CONTENTS

FIGURES

EXECUTIVE SUMMARY

1. Introduction

The Second International Mathematics Study was a comprehensive survey of the teaching and learning of mathematics in the schools of some twenty countries (educational systems) around the world. In the Study, detailed information was obtained on the content of the intended mathematics curriculum, what mathematics was actually taught by the teachers, and how that mathematics was taught. Student achievement and attitudes were assessed using internationally developed tests and questionnaires that were taken by random samples of mathematics classes in each country. The Study was targeted at 13 year olds in most countries (12 year olds in Japan and Hong Kong) and at those students at the end of secondary school who were enrolled in advanced college-preparatory mathematics courses. The lower level, younger group was called "Population A", and the older group, "Population B" (corresponding in the United States to the eighth and twelfth grades, respectively).

For each target population, topics were tested that reflected an international consensus of mathematical content judged to be important by panels of experts in each country. As a result, the fit of the tests to the curriculum varied somewhat from country to country. Data were obtained from teachers as to whether the content had been taught to the students who were tested. This information, called "opportunity-to-learn," provided a backdrop for interpreting the achievement scores.

In the United States, students in approximately 500 mathematics classrooms in about 250 public and private schools randomly selected from across the country were tested at the end of the 1981-1982 school year. (A number of countries, including the United States, also tested the students at the beginning of the school year.) Approximately 7,000 eighth grade and 5,000 twelfth grade students were involved. The other countries (systems) taking part in the Study were:

Belgium (French and Flemish)	Japan
Canada (British Columbia and Ontario)	Luxembourg†
England and Wales	Netherlands†
Finland	New Zealand
France†	Nigeria†
Hong Kong	Scotland
Hungary	Swaziland†
Israel	Sweden
	Thailand

†Tested Population A (lower secondary school) only

2. Achievement and Attitudes of U.S. Students

2.1 Population A (eighth grade in U.S. and most other countries, seventh grade in Japan and Hong Kong)

Five major topics were on the international test: arithmetic, algebra, geometry, statistics and measurement.

On all five topics, Japan obtained the highest achievement scores of all countries in the Study. Teacher coverage (opportunity-to-learn) of these topics in Japan was among the highest of any country, as well.

•••

U.S. students were slightly above the international average in computational arithmetic (calculation) and well below the international average in non-computational arithmetic (e.g. problem solving).

•••

U.S. achievement in algebra was comparable to the international average.

•••

Achievement in geometry for the U.S. was among the bottom 25 percent of all countries, reflecting to a large extent low teacher coverage of this subject matter.

• • •

Over the nearly twenty year period since the First International Mathematics Study (conducted in 1964), eighth grade U.S. mathematics classes showed a modest decline in achievement. The declines were somewhat greater for the more demanding comprehension and application items than for the computation items.

• • •

2.2 **Population B** (twelfth grade college-preparatory mathematics in the United States)

Six major topics were tested: number systems, sets and relations, algebra, geometry, elementary functions and calculus, and probability and statistics.

Hong Kong achieved the highest score on each topic, with Japan a close second. Although the Hong Kong school system is very selective, the Japanese system is not. In Japan, and several other countries (including the U.S.), between 12 and 15 percent of students were enrolled in the advanced mathematics classes.

• • •

In the U.S., the achievement of the Calculus classes, the nation's **best** *mathematics students, was at or near the* average *achievement of the advanced secondary school mathematics students in other countries. (In most countries,* **all** *advanced mathematics students take calculus. In the U.S., only about one-fifth do.)*

• • •

The achievement of the U.S. Precalculus students (the majority of twelfth grade college-preparatory mathematics students) was substantially below the international average. In some cases the U.S. ranked with the lower one-fourth of all countries in the Study, and was the lowest of the advanced industrialized countries.

• • •

Since the First International Mathematics Study, the U.S. advanced mathematics classes showed an overall modest increase in achievement, especially in functions and calculus. Much, but not all, of this improvement was due to the high performance of the Calculus classes, which constituted a greater proportion of the U.S. sample for the Second Study than for the First Study. A considerable amount of this increase was seen in the more demanding comprehension questions and, for the Calculus students, was at the even more demanding application level.

•••

2.3 Attitudes Toward Mathematics

Overall, the attitudes of U.S. students toward mathematics were positive, and compared favorably with those of other countries. Japanese students, in spite of their high achievement scores in mathematics, tended to have negative attitudes about the subject.

•••

The majority of U.S. students reported that they wanted to do well in mathematics, and that their parents wanted them to succeed.

•••

3. Opportunity-to-Learn Mathematics

The extent of coverage by U.S. mathematics teachers of topics on the international test was typically at or below the international average for most content areas, at both grade levels. Within the United States, coverage varied a great deal among classes. That is, marked differences in opportunities to learn mathematics were found between students.

•••

Even though, from an international point of view, a relatively large proportion of U.S. young people remain in school, only about 15% of these students are enrolled in advanced mathematics. This figure is at about the average for all countries. Furthermore, of these students, only a small fraction (about 3%) pursue a full calculus course. Since the achievement of these advanced students is relatively low, the mathematical yield of U.S. schools may be rated as among the lowest of any advanced industrialized country taking part in the Study.

•••

4. Curricular Differentiation and Intensity (lower secondary school)

In contrast to many other countries (most notably, Japan), the United States curriculum is dramatically differentiated at the eighth grade school level. Four mathematics programs were identified, each with vastly different mathematical content, ranging from algebra for the most able students to grade school arithmetic for the least able students.

• • •

The U.S. mathematics curriculum is characterized by a great deal of repetition and review, with the result that topics are covered with little intensity. By contrast, at this level, France places a great deal of emphasis on geometry and Japan provides an intense treatment of algebra.

• • •

The eighth grade mathematics curriculum in the U.S. tends to be "arithmetic-driven," resembling much more the end of elementary school than the beginning of high school.

• • •

5. Homework

At both grade levels in the U.S. the amount of homework reported was at, or slightly above, that reported by mathematics teachers at corresponding grade levels in other countries.

• • •

Students in about 80 percent of eighth grade mathematics classes in the U.S. were reported to spend three hours per week or less on mathematics homework. The average was estimated by teachers to be nearly 2.5 hours per week (30 minutes per day).

• • •

The typical twelfth grade U.S. mathematics student was estimated to do about four hours of homework per week.

• • •

6. Class Size

Eighth grade mathematics classes in the U.S. typically had about 26 students, which was close to the international average. By contrast, the size of mathematics classes in Japan was about 40 students.

•••

Twelfth grade U.S. mathematics classes typically had 20 students. (Again, about at the international average for this grade level.) In Japan, the average size of mathematics classes at this level was 43 students.

•••

7. Yearly Hours of Mathematics Instruction

The average amount of time per year allocated to eighth grade mathematics instruction in the U.S. was 144 hours. This compared favorably to the amount of time devoted to mathematics instruction at this grade level in the vast majority of countries in the Study.

•••

The average amount of twelfth grade mathematics instruction per year was about 150 hours per class. This was somewhat lower, on average, than the time devoted to advanced college-preparatory mathematics instruction in many countries.

•••

8. Teacher Background and Attitudes

In terms of background (amount of mathematics and mathematics pedagogy studied, years of teaching experience, and so forth) U.S. mathematics teachers do not appear markedly different from their counterparts in other countries.

•••

The teacher of U.S. eighth grade mathematics was typically experienced and well-trained, having 13 years of teaching experience, nine or ten semester courses in post-secondary school mathematics and two courses in the teaching of mathematics.

•••

The teacher of advanced high school mathematics in the U.S. had about 16 years of teaching experience with eight years at the senior level. This teacher had a median age of 40 years and had taken about 16 semester courses of post-secondary school mathematics.

• • •

Mathematics teachers of both target groups in the U.S. reported finding their classes easy to teach, with most students attentive, and relatively few especially fearful or anxious about mathematics.

• • •

Reasons given by Japanese teachers for low achievement of their mathematics classes tended to be internalized — *the teachers accepted much of the responsibility themselves. U.S. teachers, on the other hand, tended to cite* external *causes, such as lack of student ability, for their students' low achievement.*

• • •

The teaching load of U.S. mathematics teachers, from an international point of view, is high. By contrast, Japanese teachers have rather light teaching loads, especially at the advanced level.

• • •

9. Role of the Mathematics Textbook

The textbook defined "boundaries" for mathematics taught by U.S. teachers at both grade levels. Limited use was made of resources beyond the textbook for either content or teaching methods.

• • •

10. Extent of Calculator Use

Calculator usage in U.S. eighth grade mathematics classes was low (data collected in 1981-1982). Only one class in 20 used calculators for two or more periods per week. One class in three was not allowed to use calculators. Another one-third of classes reported never using calculators, even though their use was not prohibited by school policy.

• • •

More calculator use at the lower secondary school level was generally found in Europe than in the United States.

● ● ●

Calculator use in advanced (college preparatory) mathematics classes was common, not only in the U.S., but in most countries. About one-third of the U.S. advanced classes used calculators in class two or more times a week.

● ● ●

11. Recommended Steps Toward Renewal of School Mathematics in the United States

i. Restructuring of the curriculum

A fundamental revision of the U.S. school mathematics curriculum, in both form and substance, is needed. This activity should begin at the early grades of the elementary school.

● ● ●

With respect to form, the excessive repetition of topics from year to year should be eliminated. A more focused organization of the subject matter, with a more intense treatment of topics, should be considered.

● ● ●

Concerning substance, the continued dominating role of arithmetic in the junior high school curriculum results in students entering high school with very limited mathematical backgrounds. The curriculum for all students should be broadened and enriched by the inclusion of appropriate topics in geometry, probability and statistics, as well as algebra.

● ● ●

ii. Reconsideration of differentiated curricula (tracking)

The practice of the early sorting of students into curricula with vastly different content leading to substantially different opportunities to learn mathematics in high school must be carefully re-examined. At present, significant proportions of U.S. students as young as 12 or 13 years are being sorted into mathematics curricula that offer little intellectual challenge, seriously limit their chances for success in many fields of study (in high school and beyond) and greatly restrict their career choices in today's society.

• • •

iii. Role and Quality of Textbooks

In most U.S. schools, commercially published textbooks serve as the primary guides for curriculum and instruction. Any significant reform effort must take this fact into account.

• • •

iv. Mathematical Yield

The mathematical yield of a system may be thought of as the product of two quantities: the proportion of high school students that is enrolled in advanced mathematics courses and how much mathematics those students know. From an international perspective, U.S. yield in mathematics is very low. Concerted efforts need to be made to find ways to retain greater proportions of students in mathematics programs throughout high school while at the same time assisting them to achieve at higher levels.

• • •

v. Professionalization of Teaching

It is critically important that the status and rewards of teaching be significantly upgraded. Such improvement may include the reduction of teaching loads and the enhancement of the classroom environment. Professional development programs should be devised that will better provide teachers with a repertoire of strategies and knowledge that will enable them to more effectively respond to the increasing challenges of the contemporary school mathematics classroom. The effective use of technology, including calculators and computers, will undoubtedly be a significant component of such programs.

•••

THE UNDERACHIEVING CURRICULUM:

ASSESSING U.S. SCHOOL MATHEMATICS

FROM AN INTERNATIONAL PERSPECTIVE

The Nation that dramatically and boldly led the world into the age of technology is failing to provide its own children with the intellectual tools needed for the 21st century.

Educating Americans for the 21st Century.
(1983a)

Learning is the indispensable investment required for success in the "information age" we are entering. . . . The people of the United States need to know that individuals in our society who do not possess the levels of skill, literacy and training essential to this new era will be effectively disenfranchised, not simply from the material rewards that accompany competent performance, but also from the chance to participate fully in our national life.

A Nation at Risk.
(1983a)

The Second International Mathematics Study was a comprehensive survey of the teaching and learning of mathematics conducted in the early 1980's in the schools of some twenty countries around the world. In each of the countries, random samples of mathematics classes and their teachers at two grade levels (corresponding to the eighth and twelfth grades in the U.S.) were drawn nationwide. In the United States, samples of about 250 mathematics classrooms at each grade level were selected.

From a U.S. perspective, the Study was timely. While data from the hundreds of classrooms and thousands of students around the country were being analyzed, there was a flurry of national reports that announced the shortcomings of education in this country and proposed steps that should be taken in order to make needed improvements.

The present report is intended to highlight the major findings of the Second International Mathematics Study. More complete information on either the findings or on technical matters (such as sampling) is provided in the references at the end of this report.

The information from the Second International Mathematics Study reaffirms the concerns of many that mathematics education in the United States is in need of renewal. Our report is different from many others in that we offer empirical data on several aspects of school mathematics. Furthermore, through the network of national research centers of the International Association for the Evaluation of Educational Achievement, we provide a backdrop of comparative data on the teaching and learning of mathematics in other countries.

A national debate on the necessity for and shape of needed educational reform has emerged and it is in this arena of public discussion that we hope this report may play a part. The criteria for assessing the value of proposed reforms remain to be settled. We believe strongly in criteria related to the importance of effective mathematics education as essential to American economic well-being and competitiveness in the technological market place. This perspective is reflected in this report. However, the data and their implications here have import for the debate on reforms in school mathematics regardless of the criteria for designing and evaluating reform efforts.

The problems facing school mathematics, as those facing education in general, are complex. Therefore, reform measures to address those problems are likely to be complex, as well. In the Second International Mathematics Study only selected aspects of those problems were addressed. Consequently, the analyses and proposals offered here are limited to those aspects. Furthermore, the interpretations presented are those of the authors listed at the beginning of this document. We recognize that given the same data, others could reasonably draw different conclusions and implications from those offered here.

This report is written for those who are concerned about mathematics in our schools — including policy makers, curriculum developers, school board members, parents, as well as classroom teachers and administrators. We have identified what we believe to be the most important findings of the Study and point to reasons why school mathematics in the United States, from an international point of view, looks as weak as it does. We conclude by sharing some ideas as to what might be done to begin to make needed improvements.

Urbana, Illinois   Kenneth J. Travers
January 15, 1987

3

I. A TIME FOR CHANGE: TOWARD RENEWING MATHEMATICS IN U.S. SCHOOLS

The world of mathematics, science and technology has changed dramatically in the past twenty years. To assure that our students are prepared for this future, we must re-examine the content, organization, and emphases of our programs. We believe that such reexamination will justify new, strikingly different school curricula and, further, that these new curricula will attract students who find current offerings very unappealing.

J. Fey and R. Good.
(1985)

In school mathematics the United States is an underachieving nation, and our curriculum is helping to create a nation of underachievers. We are not what we ought to be; we are not even close to what we can be. It is a time for change — a time to renew school mathematics in the United States.

The Second International Mathematics Study was designed to gather data on what mathematics is taught, how it is taught and what is learned in schools of the United States and other countries. The Study was undertaken without a sense of crisis or, indeed, without an expectation of the disappointing results that emerged. Our participation in this Study sought to assess U.S. school mathematics from an international perspective so that we could evaluate our attainments and determine reasonable goals for the future.

The Study was never thought of as an international contest, a sort of "academic olympics." We were looking for answers, not winners. However, once the data became available, a sense of urgency prevailed. It became imperative to press the case for change with those concerned about school mathematics.

Our sense of urgency is not unique. But this report, unlike much current criticism of American education, is based firmly on data and inferences drawn from those data. The data-based nature of this report is reflected in the several "exhibits" that present cases for needed change.

The international mathematics tests used in the Study were developed by an international committee in cooperation with a national committee of mathematics education specialists in each country. The tests represent an international consensus as to what mathematics is expected to be taught and learned. For the United States, the curricular fit of the tests was generally satisfactory for the grade levels used in the Study.

The performance of U.S. students on the international mathematics tests was at or below the international average for the younger group (eighth grade) and was very low for the older group (twelfth grade college preparatory mathematics). In some cases, the advanced senior high school group placed among the lowest one-fourth of the nations in the Study.

Achievement is not the only important outcome of mathematics instruction. Also of importance are the attitudes and beliefs about mathematics that young people develop as they study the subject.

Every weekday, 25 million children study mathematics in our nation's schools. Those at the younger end, some 15 million of them, will enter the adult world in the period 1995-2000. The 40 classroom minutes they spend on mathematics each day are largely devoted to mastery of the computational skills which would have been needed by a shopkeeper in the year 1940 — skills needed by virtually no one today. Almost no time is spent on estimation, probability, interest, histograms, spreadsheets, or real problemsolving — things which will be commonplace in most of these young people's later lives. While the 15 million of them sit there drilling away on those arithmetic or algebra exercises, their future options are bit-by-bit eroded.

Mathematical Sciences Education Board
(1986a)

From an international perspective, U.S. students have rather neutral attitudes toward mathematics. They see it as moderately important and somewhat easy; they do not particularly like it or dislike it. The majority of our students reported that they wanted to do well in mathematics and (notably at the advanced mathematics level) were looking forward to taking more mathematics.

How can the results of the Study (especially the low achievement) be explained? Nothing could be more human than to search for responsible villains. Yet many of the commonly offered explanations are deceptive, some because they seek to identify a single culprit that explains all and leads to quick and simple solutions — others because the interplay of factors in their part of the problem is complex and easily misunderstood.

Size of class and time allocated to mathematics instruction are two deceptively simple explanations. As the Study's data show, such "easy answers" are too simple to be true. It was found, for example, that one of the highest achieving countries, Japan, has one of the largest average-size mathematics classes (about 40 students), while the U.S. is at about the international average with about 26 students at the eighth grade and 20 students at twelfth grade. In terms of time available for mathematics instruction, on the other hand, the U.S. is among the upper one-third of the countries, with about 144 hours per year devoted to the subject at the eighth grade level.

Another deceptive explanation is the argument that, since public education in the U.S. is comprehensive and available to the many and not just the few, it must deal with a more varied population of students. Therefore, we should be willing to settle for more modest levels of achievement for the larger proportions of students in our schools. Unfortunately, such an "explanation" is not supported by the data. With respect to the proportion of students who are studying advanced mathematics in school, many countries match the U.S., with between 12 and 15 percent, including, for example, Sweden and Japan. Furthermore, while in most countries all students who are enrolled in advanced mathematics study calculus, only about one-fifth of U.S. students do.

More valid explanations are concerns about the preparation and status of the teachers in our mathematics classrooms. With respect to preparation, however, U.S. teachers compare favorably with their counterparts in other countries in terms of the numbers of

The criticism of American education is properly not so much a criticism of the schools as of the society and culture generally.
For the most part, our schools have reflected those facets of our culture that are deep and abiding, and they have accordingly provided the kind of education that has been expected or demanded of them.

S. McMurrin
(1963a)

8

courses taken in mathematics and mathematics pedagogy. And there appear to be important differences of attitudes between the U.S. mathematics teachers and those of other countries (again, most notably, Japan), differences that may be accounted for in part by differences in status and working conditions. The teaching loads of U.S. mathematics teachers are among the heaviest of any of the countries in the Study. Our teachers operate in a constrained environment with scarce resources of time and energy. There is heavy dependence upon the textbook, which sets the boundaries for curricular content and instructional strategies.

There are also concerns about the nature and quality of the pedagogy demonstrated in U.S. mathematics classrooms. The data suggest that the quality of instruction could be enhanced. Little use is being made of resources beyond the textbook — even calculators and microcomputers. Only a limited repertoire of pedagogical approaches is used and those strategies emphasized often focus on the abstract and the rote. Passive reception appears to be learning style judged to be most suitable for students.

Of all of the reasons offered, however, the culprit that seems to be central to the problems of school mathematics is the curriculum. It is the mathematics curriculum that shapes the textbooks that set the boundaries of instruction. It is the mathematics curriculum that distributes goals and content during the years of schooling.

Something appears to be wrong with the way the content and goals are distributed in school mathematics in U.S. schools. Content is spread throughout the curriculum in a way that leads to very few topics being intensely pursued. Goals and expectations for learning are diffuse and unfocused. Content and goals linger from year to year so that curricula are driven and shaped by still-unmastered mathematics content begun years before.

The source of these defects in the curriculum, we conclude, is in the "spiral curriculum," as implemented in most American mathematics programs and textbooks. Conceptually elegant and intellectually appealing, the logic of the spiral curriculum has not survived the translation to the realities of classroom and market forces that shape the development of textbooks. Instead, excessive spiralling of content misdirects our mathematics curriculum. The result is a curriculum that lacks challenge and vitality.

Our educational effort has not been raised to the plateau of the age we live in. . . . We must measure it not by what it would be easy and convenient to do, but by what it is necessary to do in order that the nation may survive and flourish.

W. Lippman
(1954)

From an international perspective, our eighth grade curriculum resembles much more the end of elementary school than the beginning of secondary school. And at the twelfth grade level, many topics are dealt with only briefly, rather than a few topics being pursued in depth. Consequently, again from an international point of view, relatively few of our students are engaged in a full-fledged course in calculus and those who are so enrolled are achieving at only average levels, at best.

The mathematics curriculum, furthermore, fails to fairly distribute opportunities to learn to children. As early as the junior high school grades, tremendous differences are created in what mathematics U.S. children have the opportunity to learn and, therefore, in what they are able to achieve. These differences in opportunity set boundaries on the degree to which individual students are able to reach their fullest potential, boundaries that leave less to reward individual efforts than in any of the other countries for which data were available. Nor are these differences in opportunity-to-learn distributed appropriately so that each student receives the challenge most appropriate to her or his abilities. Socially, as well as organizationally, the mathematics curriculum in U.S. schools falls very short of its potential.

The following sections of the report set out the elements of this argument in more detail. Exhibit A gives a fuller picture of the achievement results. Exhibit B provides national patterns of opportunity-to-learn mathematics. Exhibit C describes the attitudes of U.S. students toward mathematics. Exhibit D discusses five deceptive explanations of the ills of U.S. school mathematics.

The next two sections are then devoted to a consideration of the curriculum. Section VI discusses the mathematics curriculum as a distributor of goals and content throughout schooling, while Section VII views the curriculum as a distributor of opportunities to learn mathematics.

Finally, the key findings of the Study are summarized, concluding with proposed milestones to becoming a nation of mathematical achievers rather than a nation of mathematical underachievers.

II. EXHIBIT A: THE ACHIEVEMENT OF U.S. STUDENTS

If our present curriculum were a new experimental curriculum that we were testing, we would be forced to pronounce it a failure.

Zalman Usiskin
(1985)

(Most of our country's schools have drifted into) . . . *a curriculum by default, a curriculum of minimum expectations that resists the changes needed to keep pace with the demands of preparing students for contemporary life.*

National Council of Teachers of Mathematics
(1986)

A case can be made that, by any reasonable standards, the achievement of mathematics students in U.S. schools is low. Two key points may be considered in making such a case — what standards are reasonable and how U.S. students fare by those standards.

Scores on standardized achievement and aptitude tests (such as the Scholastic Aptitude Test) have often been used to show that our achievement in mathematics, among other subjects, is not what it ought to be ("Test scores are declining . . ."). Recently these same measures have been used to announce that the crisis is past ("Test scores are rising again at last . . ."). But whether used to accuse or excuse, such information makes use of only one standard — how we are doing now as compared with our past performance. That is, we compare ourselves with ourselves.

Another yardstick that may be applied is that of absolute standards. For example, what proportion of our students have mastered a mathematical concept or skill that is judged to be an appropriate outcome for that particular group of students?

A third approach to assessing how well we are doing is to make use of international comparisons — What is education like in other countries? What mathematics do they teach, to whom and in what way? What patterns of achievement do we find across countries? And perhaps most importantly, how can we explain the profiles of high and low achievement and attitudes that we find?

We did not take part in the Study simply to find out what other countries are doing and then to imitate various promising practices. Rather, we sought information on mathematics education in other countries so that we could better understand and assess mathematics education in our own country. With this information in hand, decision-makers at various levels — national, state and local — would then be better able to take steps aimed at useful reforms in school mathematics.

In the Study, information was gathered using national samples of students from two target populations, defined internationally as Population A and Population B:

We must broaden the public's concept of what students need to learn about mathematics, while increasing performance expectations, and then have these brought to bear constructively on schools. . . . Expectations for school mathematics must be modified and raised. A national dialogue among the primary actors will be required.

Mathematical Sciences Education Board
(1986b)

Population A (Eighth Grade in the U.S.)

Population A was made up of students at the grade level in which the majority were thirteen years of age at the middle of the school year. *For the United States, this was the eighth grade.* This is an important group, since in most countries almost all children are still in school at age thirteen.

Population A is also of interest because in most countries it marks the transition from elementary to secondary schooling. Therefore, data are provided on the outcomes of mathematics education through the elementary school years. This information is important for those at the secondary school level who have the responsibility of providing appropriate programs for incoming students.

Table 1

Proportion of Population B Students in Relevant Age Groups and Grade for Each Country: 1981

Country	Age Group (Years)	Pop B Percent of Age Group	Pop B Percent of Grade Group	Percent of Age Group in School
Belgium (FL)	17	9-10	25-30	65
British Columbia	17	30	38	82
England & Wales	17	6	35	17
Finland	18	15	38	59
Hungary	17	50	100	50
Israel	17	6	10	60
Japan	17	12	13	92
New Zealand	17	11	67	17
Ontario	18	19	55	33
Scotland	16	18	42	43
Sweden	18	12	50	24
U.S.A.	17	13	15	82

Notes: 1. Age group is estimated age at middle of school year.

2. While the fourth column represents the percent of the age cohort still in school, this does not imply that all these students are in the grade(s) from which the Population B sample is drawn. Thus the second column is not always a simple product of the third and fourth columns.

3. Data are obtained from national reports for each country. The ratio of high school graduates to population age 17 was 72 percent in the United States in 1981. U.S. data on enrollment were based on the school enrollment rates of persons 17 years old according to the October 1981 Current Population Survey. An additional 5 percent were enrolled in college or university.

16

Population B (Twelfth Grade College Preparatory Mathematics in the U.S.)

This group was defined internationally as those students in their last year of secondary school who are still engaged in the serious study of mathematics. In the United States, this population consisted of students in *twelfth grade college-preparatory mathematics classes*, that is, classes requiring as prerequisites *at least* two years of algebra and one year of geometry.

Note that, in contrast with Population A, Population B is a select group. Presumably, these students are the "cream of the crop" with respect to school mathematics in the school system of each country, made up of those who intend to go on to college and, in many cases, to major in fields of study that rely heavily on a sound preparation in mathematics (for example, engineering, computer science, and physics). In order to determine how selective this "college preparatory" group of students is in each country, it was first determined what proportion of young people remains through the end of secondary school. This proportion was found to vary greatly from country to country.

The system retaining the largest proportion of young people in school is Japan, with 92% still enrolled. The lowest figures were from England and Wales, and New Zealand, with only 17% remaining in school. In the United States, 82% of our 17 year-olds are still in school — one of the highest rates of all of the countries in the Study.

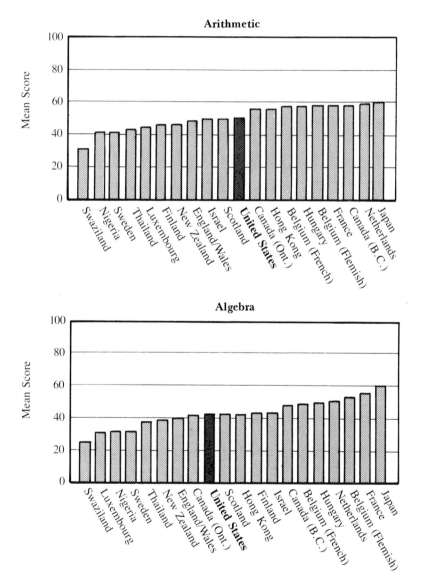

Eighth grade achievement scores in arithmetic and algebra for twenty countries show the U.S. to be in the middle group. Note, however, that countries in the middle range of achievement had very similar scores. Therefore, the ranking of those countries would be affected substantially by a change in score of only a few points.

18

For our Study, however, it is also important to know the proportion of students in school that is also in Population B (advanced college preparatory) mathematics. This figure varies greatly, too, across countries. In Hungary, for example, *all* students who remain in school are in Population B (this amounts to 50% of the seventeen year olds in the country). In Japan and the United States, about 15% are in this group.

Considerable care was taken to obtain random and representative samples of each population in each participating country. In the U.S., this involved sampling school districts across the entire country, then schools within school districts, and finally, classrooms within schools. The result was a total of about 500 classrooms from 250 public and private schools across the U.S. Data were collected from about 500 teachers (the teachers of the selected classes), 7000 eighth-grade students and 5000 students enrolled in twelfth-grade mathematics classes.*

The Study began with a comprehensive survey of the content of the mathematics curriculum in each country. With this information, the international tests were developed and pilot tested internationally. The result was a large pool of items for each population, 199 for Population A and 136 for Population B. Through a technique known as "item sampling" a large number of items could be used on the international tests in order to provide in each country national scores on a broad range of mathematics topics. Individual students, however, were required to answer only a small fraction of the items.

---

*Details of the sampling procedures used in each country are in Garden (1984). Information on U.S. sampling is also provided in the U.S. *Summary Report* (1985).

Geometry

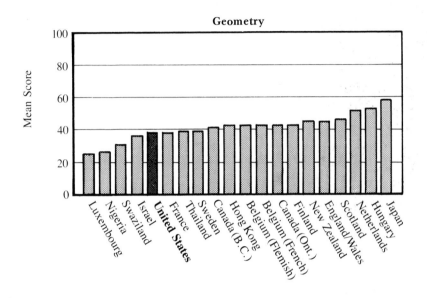

Measurement

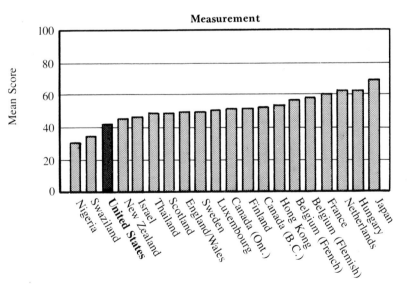

Eighth grade achievement scores in geometry and measurement for twenty countries show the U.S. to be well below the international average, among the lowest fourth of participating countries in both topics. While many measurement test items involved knowledge of the metric system, many others did not. Performance on both types of items was disappointing.

Population A Achievement:

Five content areas of mathematics were tested at this level: arithmetic, algebra, geometry, statistics and measurement. The results are given in Table 2. Notice that for three of the topics, (arithmetic, algebra and statistics) the United States scored at or near the international average. On the remaining two topics (geometry and measurement) the U.S. scores were very low — among the lowest one-quarter of the countries.

Table 2

Eighth Grade (Population A) Achievement Comparisons:
U.S. and International, 1981-82
(Percent of Items Correct)

Topic	United States (Percent Correct)	International (20 Countries) Lowest Quarter of Countries	Median	Highest Quarter of Countries
Arithmetic	51	45	51	57
Algebra	43	39	43	50
Geometry	38	38	43	45
Statistics	57	52	57	60
Measurement	42	47	51	58

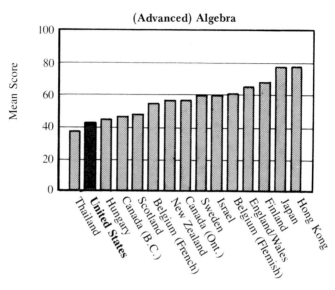

(Advanced) Algebra

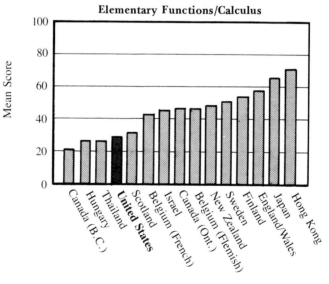

Elementary Functions/Calculus

Population B (Twelfth Grade) achievement in algebra and elementary functions/calculus for fifteen countries shows the U.S. to score among the lowest fourth of participating countries. Much scientific and technical work, in college and elsewhere, requires competence in these basic areas of mathematics.

Population B Achievement:

The results for Population B (twelfth grade college-preparatory mathematics in the United States) are given in Table 3. Six topics were tested: sets and relations, number systems, algebra, geometry, functions and calculus, and probability and statistics. The U.S. results here were more disappointing than for the eighth grade. The U.S. *never* achieved as well as the international average. On one topic (sets and relations) we were mid-way between the bottom quarter of the countries and the international average. On the remaining five topics, our scores were generally among those of the bottom one-fourth of the countries.

Table 3

Twelfth Grade (Population B) Achievement Comparisons:
U.S. and International, 1981-82
(Percent of Items Correct)

Topic	United States			International (15 Countries)		
	Pre-calculus	Cal-culus	Total	Lowest Quarter of Countries	Me-dian	Highest Quarter of Countries
Sets & Relations	54	64	56	51	61	72
Number Systems	38	48	40	40	47	59
Algebra	40	57	43	47	57	66
Geometry	30	38	31	33	42	49
Elementary Functions/Calculus	25	49	29	28	46	55
Probability/Statistics	39	48	40	38	46	64

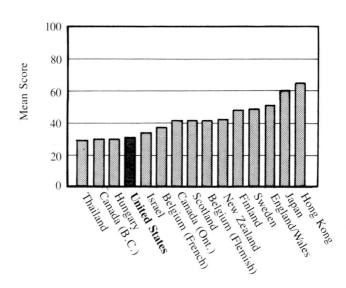

Population B (Twelfth Grade) achievement scores in geometry for fifteen countries reveal the U.S. to be among the lowest one-fourth of participating countries. Items on this subtest involved primarily knowledge of analytic geometry.

In order to better understand the U.S. findings at the twelfth grade level, our National Committee grouped the classes on the basis of the textbooks used and on supplementary information provided by classroom teachers. Two class types were identified: those pursuing an entire year of calculus and those studying a variety of advanced mathematical topics including some calculus, but less than a full year of the subject. The former were called "Calculus" classes and the latter, "Precalculus" classes. In the U.S. twelfth grade sample, only about one-fifth of the classes could be labelled "Calculus." (Here we have a striking contrast to most other countries, where, typically, all students in Population B mathematics study a full calculus course.) Our Calculus classes were therefore a very special group.

In every content area the Calculus classes had higher achievement than did the Precalculus classes. For the algebra items, the Calculus class mean was almost fifty percent higher than the mean for the Precalculus classes; for elementary functions and calculus it was almost double. The Calculus class means met or exceeded the international median for almost every topic (although never by more than a few points). But in no case did our Calculus classes approach the 75th percentile internationally. Thus, the better of these select U.S. mathematics classes still do not look very strong, by international standards. Of course, the Precalculus classes in the U.S. sample did even less well.

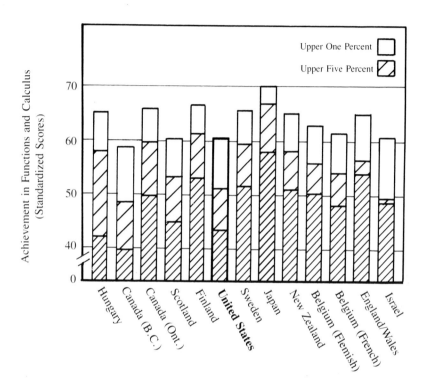

Achievement in Functions and Calculus (Standardized Scores)

Upper One Percent
Upper Five Percent

Countries (left to right): Hungary, Canada (B.C.), Canada (Ont.), Scotland, Finland, United States, Sweden, Japan, New Zealand, Belgium (Flemish), Belgium (French), England/Wales, Israel

How well the most able college preparatory mathematics students in each country achieve is shown here. The achievement of the top 1% and 5% of the students in each country as well as the average achievement for the country, are reported for functions and calculus. The most able Japanese mathematics students attained higher scores than their counterparts in other countries in both subject areas. The most able U.S. students scored the lowest of all of these countries in algebra and were among the lowest in calculus. Furthermore, as the graphs suggest, *average* Japanese students achieved higher than the top 5% of the U.S. students in college preparatory mathematics. (Countries are in decreasing order from left to right in terms of the proportion of students enrolled in advanced mathematics programs.)

Since the proportion of young people in Population B mathematics varies greatly across countries, meaningful comparisons of achievement are especially difficult for this group. One would expect a country that retains few students (presumably, the most able) until the end of secondary school to have higher achievement scores, on average, than those of a country that retains a large proportion of students in the system. In order to control this selection effect, an analysis was made of the average achievement in algebra of the top 1% and top 5% of the age group in each country. The results showed that the U.S. came out as the lowest of any country for which data were available. That is to say, the algebra achievement of our most able students (the top 1%) was lower than that of the top 1% of any other country. The algebra achievement of our top 5% group was lower than any other country, except for Israel. In functions and calculus, the achievement of the top 1% of the U.S. students exceeded Canada (British Columbia) by only a few points. But calculus is not even included in the curriculum of British Columbia!

Changes in Achievement Over the Past Twenty Years

In order to permit comparisons of mathematics achievement over the nearly twenty-year period from the First International Mathematics Study (carried out in 1964), a common set of test items was used in both studies. Table 4 reports such data for the eighth grade and Table 5 does so for the twelfth grade.

Table 4

Eighth Grade Achievement by Topic on Common Items for
1964 and 1982: U.S.

Topic (Number of Items)	1964	1982	Change
Arithmetic (14)	55	49	− 6
Algebra (10)	40	41	+ 1
Geometry (5)	40	34	− 6
Statistics (5)	57	54	− 3
Measurement (2)	35	37	+ 2
Overall (36)	48	45	− 3

Knowledge, learning, informa-
tion, and skilled intelligence are
the new raw materials of inter-
national commerce and are today
spreading throughout the world
as vigorously as miracle drugs,
synthetic fertilizers, and blue
jeans did earlier.

A Nation at Risk.
(1983c)

At the eighth grade level, the slight decline of three percentage points is surely modest. But the drop of six points for arithmetic is of concern and raises questions about the impact of the "Back to Basics Movement" of the 1970s. We should also keep in mind that our international position on these items was slightly below the mean in 1964. That is, we had a weak position twenty years ago and do not seem to have improved that position since.

For twelfth grade college-preparatory students, things look a little better on the surface, since there is an increase of six percentage points on the common items. But much of this increase was due to the improved performance on the ten functions and calculus items. We should keep in mind, however, that over the past twenty years, many more U.S. students have been taking calculus in high school (the number of students taking the Advanced Placement Calculus examination quadrupled from 8000 in 1964 to 32000 in 1982). In view of the increased enrollment, perhaps an even greater improvement in calculus scores should have been expected.

Table 5

Twelfth Grade Achievement by Topic on Common Items for 1964 and 1982: U.S.

Topics (Number of Items)	1964	1982	Change
Sets and relations (1)	21	56	+ 35
Number systems (1)	45	59	+ 14
Algebra (3)	44	45	+ 1
Geometry (4)	36	37	+ 1
Elementary functions/ calculus (10)	25	31	+ 6
Probability and statistics (1)	45	49	+ 4
Overall (20)	32	38	+ 6

Percent Of Items Taught

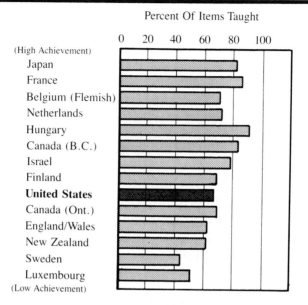

Opportunity-to-learn algebra in Population A (eighth grade in the United States) varies greatly across countries and, as the figure suggests, is related to student achievement in algebra. The countries in the figure are ranked by achievement on the international test from high (Japan) to low (Luxembourg). For each country, the length of the bar shows the opportunity-to-learn score for algebra. That is, we are shown the average number of algebra items on the international test whose mathematical content was reported by the classes' mathematics teachers to have been taught either before or during the school year.

High achieving countries, such as Japan and France, had high opportunity-to-learn scores for algebra, with teachers stating that the content of over 80 percent of the items on the international test had been taught to their students. Teachers in low achieving countries, such as Sweden and Luxembourg, reported having taught only about one-half of the algebra items on the test.

In the United States, teachers reported that nearly 70 percent of the algebra on the international test had been taught. This is slightly below the international average for opportunity-to-learn, as the lengths of the bars suggest. (Opportunity-to-learn data were not available for all countries.)

Opportunity-to-learn geometry in Population A was relatively low in all countries, except for Hungary. On average, only about one-half of the content of the geometry items on the international test was taught. In several countries, including the United States, about 40 percent or less was taught. There is little agreement among countries as to what topics in geometry should be in the curriculum for this population of students (eighth grade in the United States).

A novel and very important feature of the Second International Mathematics Study was that of obtaining reports from the teachers of the classes on whether the mathematics on the test had been taught, either during the year or prior to that year. The resulting data, called "opportunity-to-learn," provide useful information for interpreting achievement scores both from country to country and within a particular national system.

Population A: Opportunity-to-Learn

The countries with the highest ratings from teachers on how much mathematics was taught were Hungary and Japan. In arithmetic and algebra, France and British Columbia reported high ratings, while Sweden was very low on these topics. (See figure *Opportunity-to-Learn Algebra...*) For most topics, the United States was at about the international average, with rather low opportunity-to-learn scores for geometry. Reasons for these "low to middling" ratings for opportunity-to-learn mathematics in U.S. classrooms are examined later in this report. (See Section VII, "The curriculum as distributor of opportunity-to-learn.")

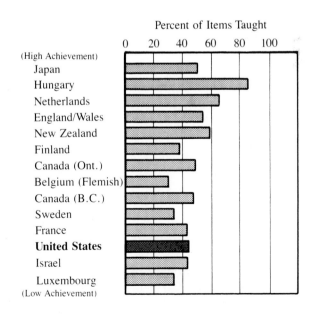

Percent of Items Taught

```
                        0    20   40   60   80   100
(High Achievement)
Japan
Hungary
Netherlands
England/Wales
New Zealand
Finland
Canada (Ont.)
Belgium (Flemish)
Canada (B.C.)
Sweden
France
United States
Israel
Luxembourg
(Low Achievement)
```

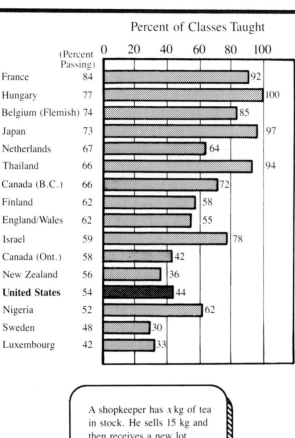

Percent of Classes Taught

	(Percent Passing)	Percent of Classes Taught
France	84	92
Hungary	77	100
Belgium (Flemish)	74	85
Japan	73	97
Netherlands	67	64
Thailand	66	94
Canada (B.C.)	66	72
Finland	62	58
England/Wales	62	55
Israel	59	78
Canada (Ont.)	58	42
New Zealand	56	36
United States	54	44
Nigeria	52	62
Sweden	48	30
Luxembourg	42	33

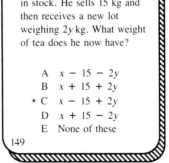

A shopkeeper has x kg of tea in stock. He sells 15 kg and then receives a new lot weighing $2y$ kg. What weight of tea does he now have?

A $x - 15 - 2y$
B $x + 15 + 2y$
★ C $x - 15 + 2y$
D $x + 15 - 2y$
E None of these

149

Expressing a word problem algebraically, as tested by this item, was taught to over 90% of the Population A classes in France, Hungary and Japan. The high achievement (over 70%) on this item in those countries reflects high opportunity-to-learn the mathematics of this item.

Internationally, the opportunity-to-learn data correspond rather closely to the achievement data, as would be expected. In Japan, opportunity-to-learn ratings are among the highest of any country. In the United States these ratings are about at the international average, or in some cases lower. This was true for both Populations A and B.

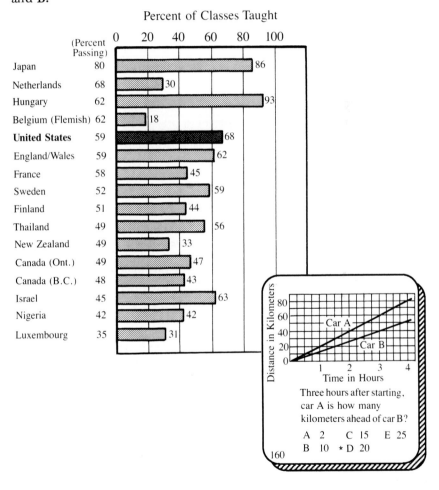

Percent of Classes Taught

Three hours after starting, car A is how many kilometers ahead of car B?

A 2   C 15   E 25
B 10   ★ D 20

Interpreting graphs, as represented by this item, was taught to nearly 70% of the eighth grade mathematics classes in the United States, with correspondingly high achievement.

Percent of Items Taught

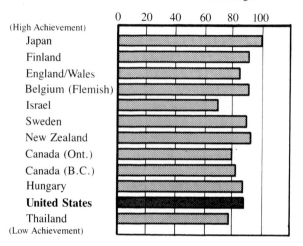

Opportunity-to-learn advanced algebra is, on average, high for college preparatory mathematics classes in all countries.

Percent of Items Taught

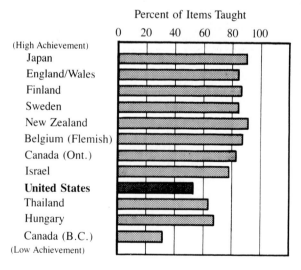

Opportunity-to-learn elementary functions and calculus in advanced high school mathematics classes varies a great deal among countries. In Japan, nearly all of the mathematical content of the functions and calculus items was covered by the teachers. However, in the United States, only about one-half of this mathematics was reported to have been taught.

Population B: Opportunity-to-Learn

At the senior secondary school level, Japan had opportunity-to-learn ratings that were among the highest of all countries (note that Hong Kong did not report this information — but their even higher achievement than Japan suggests very high opportunity-to-learn scores for them as well). In fact, in advanced algebra, 100% of the Japanese teachers reported coverage of this topic, while in functions and calculus, the opportunity score for Japan was 94%. By contrast, in the United States, the opportunity-to-learn ratings were 89% for algebra and only 58% for functions and calculus.

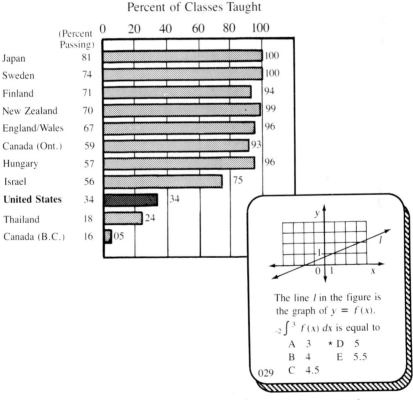

Percent of Classes Taught

An understanding of calculus (integration) is needed to correctly answer this item. In most countries, the advanced senior secondary school mathematics classes are taught calculus, as the high opportunity-to-learn scores suggest. In the United States, however, only a small proportion of twelfth grade college preparatory mathematics students study the subject.

IV. EXHIBIT C. THE ATTITUDES OF U.S. STUDENTS

(In the National Assessment of Educational Progress) . . . Nine-year-olds rated mathematics as the best liked of five academic subjects; 13-year-olds rated it as the second best liked subject; and 17-year-olds rated it as the least liked subject.

T. Carpenter, et al.
(1981)

Achievement in mathematics is not the only important outcome of instruction. Also of serious concern are the attitudes and beliefs that young people develop about mathematics.

The Study made use of a large number of items that surveyed student (and teacher) attitudes and beliefs both about the specific activities in which they participated in mathematics classes and about their conception of mathematics more generally. These items were also clustered into several scales of related items. (See, for example, the figure *Activities in the lower secondary school. . .*).

Since the same items were used for both populations, interesting contrasts of the attitudes and opinions of the eighth and twelfth graders were possible. The eighth-graders represented a more general cohort of students who had not yet specialized in mathematics. The eighth-grade students were those who were approaching the years when they must make their first decisions as to whether to pursue the study of mathematics. In viewing these findings, it is important to keep in mind that the twelfth-grade population represented a group of specialists with a demonstrated interest in studying mathematics.

It is, of course, perfectly possible to develop appropriate attitudes and beliefs among those attracted to and pursuing mathematics learning while at the same time failing to develop attitudes and beliefs among the more general run of students. The contrast between attitudes of eighth- and twelfth-grade students can help determine if this is the case.

The attitude items of the Study provided data on two critical, areas of belief — student perceptions of the importance of mathematics and their commitment to it, and student conceptions of the nature of mathematics and their reactions to it.

Mathematics In School†
(Population A, Eighth Grade in U.S.)

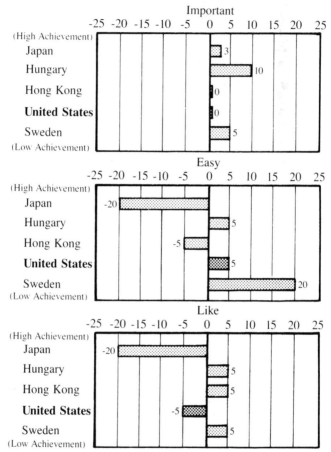

Important

| | -25 | -20 | -15 | -10 | -5 | 0 | 5 | 10 | 15 | 20 | 25 |

(High Achievement)
Japan — 3
Hungary — 10
Hong Kong — 0
United States — 0
Sweden — 5
(Low Achievement)

Easy

(High Achievement)
Japan — -20
Hungary — 5
Hong Kong — -5
United States — 5
Sweden — 20
(Low Achievement)

Like

(High Achievement)
Japan — -20
Hungary — 5
Hong Kong — 5
United States — -5
Sweden — 5
(Low Achievement)

Activities in the lower secondary school (Population A) mathematics classroom were rated by the students as to importance, ease and whether they were liked. The activities included memorizing rules and formulae, checking answers to problems and drawing geometric figures. As the above data indicate, the Hungarian students found these mathematics activities rather important, as did the Swedish students. The Swedish students, as well, found the activities very easy. The Japanese students found the activities very difficult and did not like them. The U.S. students were, comparatively, somewhat neutral in their attitudes toward these mathematics classroom activities.

†In the above figure, as well as in the next one (**Mathematics in School, Population B**), the data reflect country differences from overall average scores as determined by the Median Polish procedure. For more details, see Kifer and Robitaille (1985).

Mathematics in School:

Students at each of the grade levels were asked to rate selected school mathematics activities on their importance, their difficulty and the degree to which they were enjoyable. For eighth-grade students in the U.S., the mathematics classroom activity rated important by the largest percentage of students was memorizing rules and formulae. The activities rated as least important were those of drawing geometric figures and using calculators. Far and away, the activity considered the most difficult and least enjoyable was memorizing rules and formulae. That most often considered easy and enjoyable was using calculators. The twelfth graders also saw calculator use as more important and memorizing as being somewhat less important.

These data are in part confirmed by other information on how mathematics was taught to these students — data showing that instruction stressed presenting rules and formulae, and in which only a minority were allowed to make use of calculators in the classroom (especially at the eighth grade).

Mathematics in School
(Population B, Twelfth Grade College
Preparatory Math in U.S.)

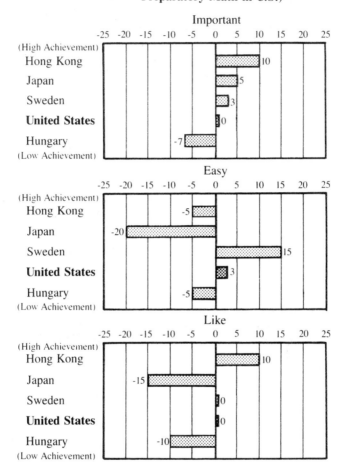

Important

| | -25 | -20 | -15 | -10 | -5 | 0 | 5 | 10 | 15 | 20 | 25 |

(High Achievement)
Hong Kong — 10
Japan — 5
Sweden — 3
United States — 0
Hungary — -7
(Low Achievement)

Easy

(High Achievement)
Hong Kong — -5
Japan — -20
Sweden — 15
United States — 3
Hungary — -5
(Low Achievement)

Like

(High Achievement)
Hong Kong — 10
Japan — -15
Sweden — 0
United States — 0
Hungary — -10
(Low Achievement)

Activities in the last year of secondary school (Population B) mathematics classroom were also rated as to importance, ease and whether they were liked. These activities included checking answers to problems, memorizing rules and formulae and proving theorems. As the above data indicate, the students in Hong Kong found these activities rather important, as did the Japanese. The Swedish students reported them to be easy. The Japanese found these activities hard and did not like them. Again, the U.S. students were, comparatively, some-what "middle of the road" in their attitudes toward these activities.

Internationally, U.S. students are seen as having less marked views about mathematics in school. For example, students in Hong Kong and Japan believed the mathematics classroom activities to be more important than did their U.S. counterparts, but these students also found the activities to be more difficult. Swedish students found the mathematics activities much easier than did the U.S. students. Overall, Japanese students had a greater dislike for mathematics activities than did students in other countries, including the United States.

Checking Work

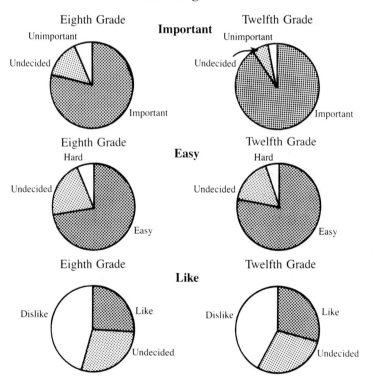

Checking your work in mathematics class was recognized as important by the vast majority of students at both grade levels in the United States. Almost as many students agreed that this was easy to do. However, relatively few students like to check their work — in fact, nearly one-half of them reported disliking it.

Memorizing Rules and Formulae

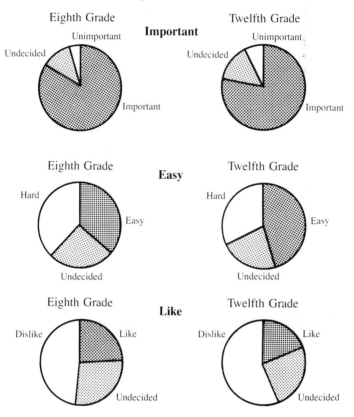

A conception of mathematics as mostly memorizing was revealed in part by eighth grade U.S. students' ratings of the activity of "memorizing rules and formulae." Almost all considered this activity to be important although only a minority found it easy and relatively few liked it. This was in marked contrast to their perception of the activity "using calculators."

Mathematics as memorized seems to have been a viewpoint shared even by the twelfth grade "mathematics specialist" student in the U.S. They too indicated the activity of "memorizing rules and formulae" as one widely seen as important, thought by less than one-half of the students to be easy, and liked by fewer than one-fourth of these students.

Using Calculators

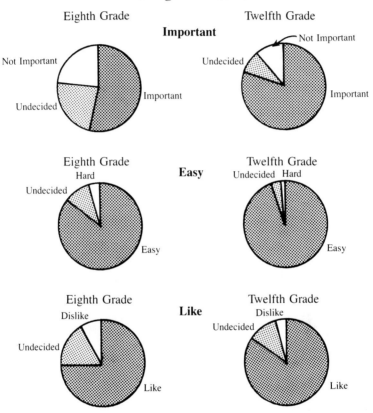

Using calculators was the best liked school mathematics activity by eighth grade U.S. students and that most widely considered to be easy. But on the other hand, using calculators was considered important by a smaller percentage of students than any other activity listed.

The importance of using calculators was much more widely accepted by the twelfth grade "mathematics specialist" students in the United States. They, too, liked calculators and found their use easy. The differing perceptions of the importance of using calculators displayed by eighth and twelfth grade students may be explained by the twelfth grade students' more widespread use of calculators and use in activities more central to mathematics instruction.

It is important to know Mathematics in order to get a good job.

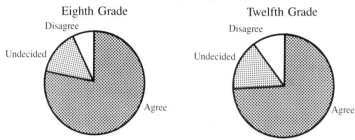

Eighth Grade

Disagree

Undecided

Agree

Twelfth Grade

Disagree

Undecided

Agree

I would like to work at a job that lets me use Mathematics.

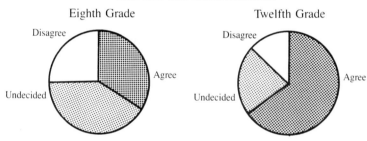

Eighth Grade

Disagree

Agree

Undecided

Twelfth Grade

Disagree

Agree

Undecided

The **importance of knowing mathematics** to get a good job was recognized by most U.S. students at both grade levels. The open question remained: did these same students perceive mathematics as important for their personal futures?

The desire to use mathematics in their future careers was expressed by a majority of U.S. students in the more selective twelfth grade population and by just over one-fourth of the eighth grade students. In spite of the shared perception of the importance of mathematics in the world of work, relatively few of the younger students seemed inclined to pursue it in their own future careers. Unless this belief is changed in the intervening years, the majority of the eighth grade students are likely to be among those who avoid mathematics courses before the end of high school.

Mathematics and Society

Student perceptions of the importance of mathematics were explored through a scale on the importance of mathematics in society. Some of the most positive responses to any of the items were obtained on this scale. Both eighth- and twelfth-grade students in the United States indicated that it was important to know mathematics in order to get a good job, that a knowledge of mathematics was necessary in most occupations and that it was useful and needed in everyday life. About one-third of the eighth-grade students and two-thirds of the twelfth-grade advanced mathematics students indicated that they would like to work at a job that used mathematics.

I really want to do well in Mathematics.

Eighth Grade

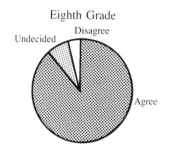

Twelfth Grade

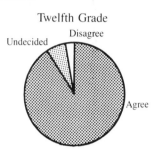

I am looking forward to taking more Mathematics.

Eighth Grade

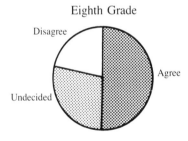

Twelfth Grade

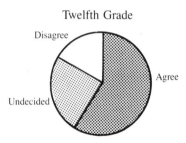

The desire to do well in mathematics was expressed by almost all U.S. students sampled at both grade levels. Still at issue: did this desire to do well extend to a willingness to take more mathematics courses?

Taking more mathematics was looked forward to by only about one-half of the U.S. students at each grade level, in spite of the widely indicated desire to do well in mathematics. Less than one-half of the eighth grade students looked forward to taking mathematics in high school. Of more concern, just over one-half of the twelfth grade "mathematics specialist" students looked forward to taking mathematics in college. Many, of course, were likely to continue work in mathematics even though they did not look forward to it.

Mathematics and Myself

Another set of items explored students' perceptions of themselves as learners and users of mathematics. Most responses indicated that the U.S. students had a somewhat positive perception of themselves as mathematics learners. A clear majority of students at both grade levels in the U.S. agreed that they really wanted to do well in mathematics (and that their parents also wanted them to do well), that they felt good when they solved a mathematics problem by themselves and that they usually understood what was being talked about in class. Also, most indicated that they didn't believe mathematics was harder for them than for most people and it did not frighten them to take mathematics.

Yet less than one-half of the eighth-grade students and only about sixty percent of the twelfth-grade college preparatory mathematics students believed that mathematics was enjoyable and about the same proportions reported that they were looking forward to taking more mathematics. Less than one-half of the eighth-graders and just over one-half of the twelfth-graders indicated that they did not spend a lot of their own time doing mathematics. Slightly more than one-half of the eighth-graders and slightly less than one-half of the twelfth-graders agreed that they would work a long time in order to understand a new idea in mathematics.

In this light, the students perceive themselves as capable in mathematics but with a limited commitment to it. Mathematics was something in which they saw themselves as able to do well and from which they gained some satisfaction. Yet it was something that they did not really perceive as enjoyable and that they did not look forward to studying.

The same pattern of perceived importance, with limited personal commitment to mathematics, was seen both in student views of the place of mathematics in society and in their views of mathematics' place in their own lives. This was true for a majority of the eighth-grade students and for a distressingly large proportion of the twelfth-grade mathematics "specialist" students, as well.

Mathematics is a good field for creative people.

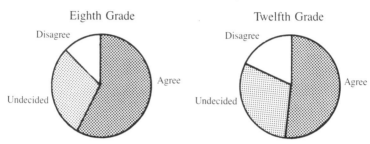

Eighth Grade   Twelfth Grade

New discoveries in Mathematics are constantly being made.

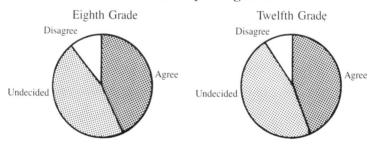

Eighth Grade   Twelfth Grade

Mathematics as a good field for creative people was a point of view accepted by just over one-half of the U.S. students at both grade levels. U.S. students' conceptions of mathematics seemed not to emphasize its creative side. This may well have contributed to a lack of commitment to learning more mathematics or to use it vocationally.

Mathematics as a changing field was an idea not accepted by many U.S. students at either grade level. Less than one-half agreed that new discoveries in mathematics were constantly being made. Many of their teachers did not agree with this statement, either. Mathematics was more often seen by students as a body of rules to be memorized and applied.

How Mathematics is Viewed as a Subject
(Mathematics as a Process)

These items explored whether students conceived of mathematics as primarily a matter of "process" rather than of "product" — as a growing rather than fixed body of knowledge and as a place for creativity and originality rather than only as a place for passive absorption of knowledge.

Forty percent or more of both grade levels in the U.S. agreed with the statement that mathematics is a set of rules. Almost one-half of the eighth-graders and about one-fourth of the twelfth-grade mathematics students believed that learning mathematics involves mostly memorizing. Over eighty percent of the eighth-graders and two-thirds of the twelfth-graders gave the opinion that there is always a rule to follow in solving a mathematics problem. About one-half of the eighth-grade students and three-fourths of those in the twelfth grade agreed that trial and error can often be used to solve a mathematics problem.

Overall, the U.S. students, especially those in the eighth grade, regarded mathematics as a largely fixed, rule-oriented body of knowledge to be acquired through memorizing and learning to apply rules. Mathematics was not widely seen as a subject in which originality or creativity may play important roles. Clearly, this conception of mathematics is different from that of those who make use of mathematics professionally. To the extent that this point of view discourages students from pursuing mathematics further, it is a harmful misconception.

V. EXHIBIT D: FIVE DECEPTIVE EXPLANATIONS

Nothing is so tempting, so frustrating, and so characteristically human as the search for villains and heroes.

H. S. Broudy
(1972)

It is true that complex enterprises generate complex problems requiring equally complex solutions. Schooling is such an enterprise. Therefore, solutions to the problems of schooling must, inevitably, be complex.

Yet solutions in a public enterprise such as schooling are, typically, pragmatically difficult, economically draining and politically dangerous. It is perhaps these difficulties that turn the "characteristically human" search for "villains and heroes" toward a longing for a simple definitive villain that can be overcome by a clear, effective hero — that is, for one or a few causes of a complex problem that would be amenable to a clear, practical solution.

The longing for simplicity in the face of essential complexity is likely to produce deceptive explanations that lead to ineffective solutions. This phenomenon has been clearly seen in recent efforts to diagnose and prescribe remedies for the failures of American public education.

Included among the factors recently identified as single causes of deficiencies in student achievement in U.S. schools have been the amount of time allocated for instruction, the size of classes, the comprehensive nature of American public schooling, the preparation and status of teachers, and the soundness of the teaching provided.

While it is likely that any or all of these factors may be parts of the complex problems of schooling in America, it is even more likely to be misleading, no matter how tempting, to single out any one of these factors as *the* villain — *the* thing to be remedied. International data from the Study suggest that such simplistic explanations are deceptive, at least with respect to mathematics education.

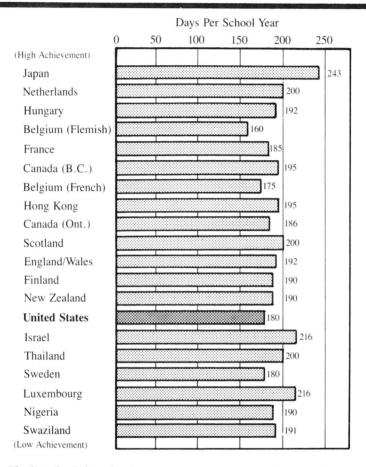

Days Per School Year

	Days
(High Achievement)	
Japan	243
Netherlands	200
Hungary	192
Belgium (Flemish)	160
France	185
Canada (B.C.)	195
Belgium (French)	175
Hong Kong	195
Canada (Ont.)	186
Scotland	200
England/Wales	192
Finland	190
New Zealand	190
United States	180
Israel	216
Thailand	200
Sweden	180
Luxembourg	216
Nigeria	190
Swaziland	191
(Low Achievement)	

The length of the school year appears to show why Japan led in mathematics achievement at the Population A level. Countries are arranged in order of average score from high to low on the international test. The lack of a systematic relationship between length of school year in days and achievement can be inferred from the absence of an orderly pattern of decreasing lengths of school years from top to bottom.

Yearly hours of mathematics instruction at Population A provide a pattern that differs from the length of the school year. For example, Japan has the longest school year but has one of the least amounts of school time devoted to mathematics. (Many Japanese students spend a considerable amount of time studying mathematics outside of regular school hours, however.) In the United States, the school year is relatively short, but a relatively large amount of that time is allocated to mathematics.

Deceptive Explanation Number One: Time for Mathematics Instruction

One popular candidate for explaining low achievement is the lack of sufficient time allocated for instruction — students learn primarily what they are taught, teaching takes time, and additional time devoted to instruction can reasonably be expected to result in gains in achievement.

Yet data from the Study indicate that the relationship between instructional time and achievement is neither simple nor linear. As the graph of our data indicates (see *Yearly hours of mathematics instruction*) some countries with *high* average achievement on the international test have a relatively *small* amount of time per year allocated to mathematics instruction (compare, for example, the U.S. and Japan). The data in the graph are for Population A, but similar findings were obtained for Population B.

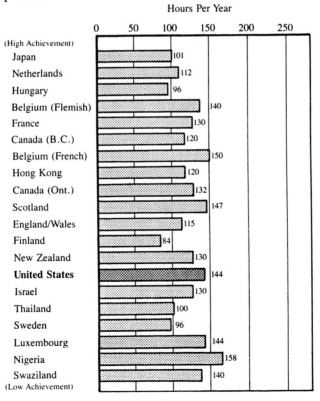

Hours Per Year

(High Achievement) Japan	101
Netherlands	112
Hungary	96
Belgium (Flemish)	140
France	130
Canada (B.C.)	120
Belgium (French)	150
Hong Kong	120
Canada (Ont.)	132
Scotland	147
England/Wales	115
Finland	84
New Zealand	130
United States	144
Israel	130
Thailand	100
Sweden	96
Luxembourg	144
Nigeria	158
Swaziland (Low Achievement)	140

With all our respect for the worth of education, we cannot be sure that our society's commitment of its human and material resources to the tasks of the schools will be adequate to satisfy our rapidly expanding need for highly developed intelligence, skill and creative talent.

S. McMurrin
(1963b)

While time devoted to instruction certainly plays an important role as one factor among many in determining levels of achievement, it is not a clear candidate for a sole, sufficient explanation. For example, the quality of instruction and of the preparation of the students receiving instruction must be considered.

Deceptive Explanation Number Two: Class Size

Class size is another favorite candidate to be offered as a cause of low achievement. Certainly, common sense suggests that the same amount of time devoted to teaching a larger class and for a smaller class will result in less time devoted to each individual in the larger class as well as a greater demand for the more complex skills of large group instruction. Large classes typically generate additional needs to "manage" the flow of activities in the classroom in a way that will help students to remain engaged in learning. And if differences in class size are sufficiently great, they can affect which pedagogical strategies are feasible.

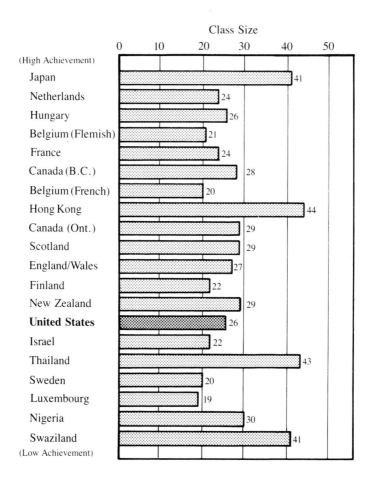

Class Size

	(High Achievement)	
Japan	41
Netherlands	24
Hungary	26
Belgium (Flemish)	21
France	24
Canada (B.C.)	28
Belgium (French)	20
Hong Kong	44
Canada (Ont.)	29
Scotland	29
England/Wales	27
Finland	22
New Zealand	29
United States	26
Israel	22
Thailand	43
Sweden	20
Luxembourg	19
Nigeria	30
Swaziland	41
	(Low Achievement)	

Class size is a deceptive explanation of level of achievement, as can be seen from this presentation of average class sizes (with countries arranged in order by achievement) for Population A. The four countries with largest classes are distributed throughout the ordered list of twenty countries, as are the four countries with the smallest class sizes.

Yet, in spite of these potential effects of class size, is it reasonable to fasten on this factor as *the* primary explanation for differences in achievement? Comparisons among participating countries again suggest that the answer is "no."

The figures *Class Size* (for Population A) and *Average size of advanced mathematics classes* (for Population B) present data to help address this issue. In both figures, countries are ordered from high to low on overall achievement on the international test. Again, no simple relationship between class size and achievement level is evidenced.

It is likely that class size, among other factors, plays a significant role in determining quality of instruction and the level of student achievement. However, it is not reasonable on the basis of the Study data to view class size as a sole explanation for levels of achievement.

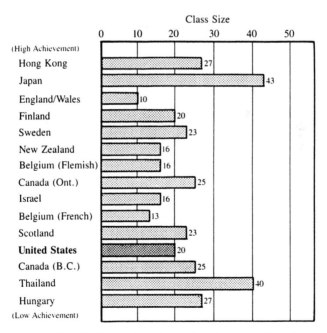

Class Size

The average size of advanced mathematics classes tends to be smaller than those of Population A, although they remain large in Japan and Thailand. As in Population A, there appears to be no relationship between a country's size of class and its ranking in mathematics achievement.

We have scarcely begun the task of providing universal education through the secondary school — that great experiment for which this century has been praised.

J. I. Goodlad
(1984a)

A high level of shared education is essential to a free, democratic society and to the fostering of a common culture, especially in a country that prides itself on pluralism and individual freedom.

A Nation at Risk
(1983d)

Deceptive Explanation Number Three: Comprehensive Education

Historically, public education in the United States has been comprehensive; that is, it has been committed to providing a sound, basic, education for all students, rather than the select few. Therefore, over the years, the U.S. has had a tradition of enrolling proportionately large numbers of children through to the end of secondary school. This contrasts, for example, with European education which, until recently, has been very selective in terms of the proportion of students completing secondary school.

The numbers of students retained in school may be regarded as a candidate to help explain low U.S. achievement. If we retain proportionately more children than do most other countries, our average test scores would be expected to be lower than those of more selective systems. So, in order to explore this idea a little more, let us look again at retention data for the countries in the Study (see Table 1, page 16). We see, for example, that in Hungary, 50% of their young people remain in school while in New Zealand and England, retention is only 17%. In the United States, 82% of our youth remain in school, a figure considerably *less* than that of the Japanese (92%).

The Japanese retention figure is even more impressive in that at the time of the First International Mathematics Study, only 57% of their youngsters completed high school. So, Japan has been able to bring about a dramatic growth rate in retentivity while at the same time keeping their very high international standing in terms of mathematics achievement! During the same time period, 1964 to 1982, U.S. retentivity increased by only a few points.

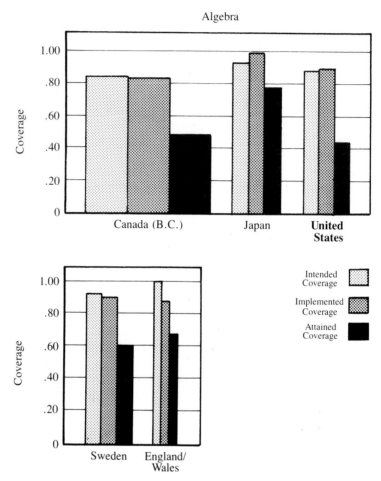

Algebra

Advanced topics in algebra are studied in the senior secondary school "mathematics specialist" classes in all countries in the Study. The above figure shows that for five countries, well over 80% of content of the algebra items on the international test was intended to be taught and about the same proportion was reported to be taught (implemented) by the classroom teachers in each country. Achievement in algebra (attained curriculum) was high in Japan and low in the United States. The width of the bars reflects the proportion of young people in each country that is enrolled in senior secondary school mathematics. Canada (British Columbia) has a high proportion (30%) while England and Wales enroll relatively few (6%). The proportions enrolled in the most advanced mathematics classes in Sweden, Japan and the United States are very similar — between 12% and 15%.

We must keep in mind, however, that at the advanced level our Study tested a special group — therefore, when looking at performance across countries, we really should be looking at the selectivity of this college-preparatory group. Table 1 provided us with this information. In the U.S., about 13 percent of 17-year-olds are enrolled in advanced (Population B) mathematics. But we also note that several other countries enrolled a similar proportion of their young people in this level of mathematics (for example, New Zealand, Sweden and Japan). It is *not* the case, therefore, that for advanced mathematics, the U.S. enrolls a proportionately larger proportion of students than other countries. In fact, we rank about average in terms of the proportion of students who take advanced mathematics courses.

Retention of students in advanced mathematics raises a related concern — that of the mathematical "yield" or productivity of the educational system. This yield may be defined in terms of two factors: (1) the proportion of young people in advanced mathematics courses and (2) how much mathematics those students learn. Mathematical yield is the product of these two quantities — a measure of "how many students know how much mathematics."

The figure *Advanced topics in algebra . .* shows that British Columbia has a high yield in algebra. That is, a relatively high proportion of their young people are enrolled in advanced mathematics and those students achieve rather well in algebra. English and Welsh students also achieve well, in algebra, but only a rather small proportion of students are studying the subject. In calculus, too, students in England and Wales achieve fairly well — but again, only a small proportion of students take these classes. In the United States, the proportion of students who study advanced mathematics is about average and the students in those courses have low achievement scores. Japan, by contrast, is a high yield country — the result of an average retention rate coupled with very high achievement.

*If they (the schools) are
failing, as many people think
they are, it is not because
teachers should have 80
semesters rather than fifty
in English literature — or because they use the wrong one
of several possible methods . . .
(Schools) are failing because
they are malfunctioning social
systems, more preoccupied
with maintaining their daily
routines and regularities than
with creating a setting where
human beings will learn and
live together productively
and harmoniously.*

J. I. Goodlad
(1975)

Deceptive Explanation Number Four: Preparation and Status of Teachers

Thus far, the "explanations" considered have focused on the teaching situation or on the educational system more generally. Another set of factors relates to those who provide the instruction. What teachers are like, how they are regarded and what they do have a major impact on education, becoming prime candidates for explanations of levels of achievement. The Study was designed primarily to explore the nature of mathematics instruction and its effects on student attainment. Information gathered on teacher preparation and status was incidental to the main thrusts of the Study and thus only a few data-based inferences are drawn about them.

It sometimes has been asserted that the limited attainments of U.S. students can be explained by the background of U.S. mathematics teachers — their lack of appropriate training and experience. The Study data do not support such an assertion. In fact, the teachers who provided instruction to the sampled classrooms were well-trained and experienced. If they were atypical (although it is believed that they were not), the findings of the Study become even more disturbing since they suggest that the limited levels of mathematics achievement seen in the sampled classes were produced by instruction provided by atypically well-trained and experienced teachers. The instruction provided by typical (less well qualified) teachers would then be expected to lead to even lower levels of achievement.

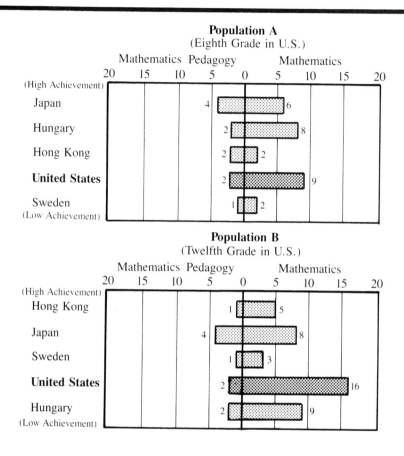

Population A
(Eighth Grade in U.S.)

Mathematics Pedagogy   Mathematics

| 20 | 15 | 10 | 5 | 0 | 5 | 10 | 15 | 20 |

(High Achievement)

Japan — 4 | 6

Hungary — 2 | 8

Hong Kong — 2 | 2

United States — 2 | 9

Sweden — 1 | 2
(Low Achievement)

Population B
(Twelfth Grade in U.S.)

Mathematics Pedagogy   Mathematics

| 20 | 15 | 10 | 5 | 0 | 5 | 10 | 15 | 20 |

(High Achievement)
Hong Kong — 1 | 5

Japan — 4 | 8

Sweden — 1 | 3

United States — 2 | 16

Hungary — 2 | 9
(Low Achievement)

    The number of courses in mathematics and in mathematics pedagogy (teaching of mathematics) taken by mathematics teachers is shown here. U.S. mathematics teachers at both the eighth and twelfth grade levels compare favorably with those of other countries. On average, U.S. eighth grade mathematics teachers report having taken nine semester courses in mathematics beyond high school, compared with six semester courses in Japan and only two in each of Hong Kong and Sweden. Similar patterns are found for teachers of advanced secondary school mathematics. U.S. teachers report having taken about as many courses in the teaching of mathematics as their counterparts in Hong Kong and Hungary, but only about one-half as much coursework in mathematics pedagogy as Japanese teachers.

64

According to our data, U.S. eighth grade (Population A) mathematics classes typically had a teacher who was about 37 years old and with 13 years of teaching experience, eight of them in teaching eighth grade mathematics. This teacher's collegiate training included nine or ten semester courses in mathematics and two courses in the teaching of mathematics. Relatively few younger teachers (only 15 percent below the age of 30), few inexperienced teachers (only 15 percent with less than seven years experience), and few teachers with limited training in mathematics (less than 20 percent reported having fewer than five semester courses of post-secondary mathematics) were found in the eighth grade sample.

At the twelfth grade level, the mathematics teacher typically was 40 years old with almost 16 years of teaching experience, eight of these teaching twelfth grade mathematics. About ten percent of the teachers were between 55 and 65 years of age. The background of these teachers included a median of 16 semester courses in mathematics and two courses in the teaching of mathematics.

In terms of the number of collegiate courses taken, U.S. teachers compared favorably with those of other countries participating in the Study, even those from high-achieving countries such as Japan. One difference noted was that Japanese teachers had somewhat fewer courses in mathematics content but more in the pedagogy of mathematics than did their U.S. counterparts.

While the levels of training and experience of U.S. mathematics teachers were not markedly dissimilar from those of teachers in high-achieving countries, some major differences were noted in the attitudes of teachers. For example, the U.S. teachers reported that mathematics was rather easy to teach. The Japanese stated that it was difficult to teach.

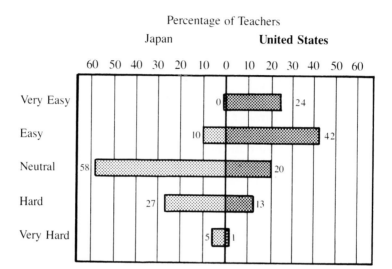

Percentage of Teachers

Japan   **United States**

| | 60 | 50 | 40 | 30 | 20 | 10 | 0 | 10 | 20 | 30 | 40 | 50 | 60 |

Very Easy   0   24

Easy   10   42

Neutral   58   20

Hard   27   13

Very Hard   5   1

How easy is it to teach mathematics? Mathematics was perceived by U.S. eighth grade teachers to be easy to teach. In Japan, the subject was regarded as more difficult to teach.

The teachers were also asked to give reasons for the lack of desired progress of their students in mathematics. The U.S. teachers tended to offer reasons related to students — student misbehavior, lack of attention or attendance, etc. — while Japanese teachers tended to attribute lack of student achievement to their own professional limitations.

Taken together, these data suggest that Japanese teachers perceived teaching mathematics as a difficult, demanding enterprise, the success of which had considerable impact on the achievement of their students. By contrast, U.S. teachers seemed to see teaching mathematics as less demanding and to view the learning of mathematics as an enterprise over which they had relatively little control.

There were also some indications, although less direct, that teachers in the U.S. and Japan were given somewhat differing status. Let us assume, for example, that teaching load may be taken as an indicator of the status attributed to teachers as well as a major factor affecting the quality of the teachers' working environments.

From an international point of view, the teaching loads of U.S. mathematics teachers are heavy. In terms of hours per week, the average assignment for the U.S. teacher of 23 hours per week for eighth grade teachers was exceeded only by the Netherlands, with 24 hours. The majority of the countries have teaching loads well below twenty hours per week, with Flemish Belgium and Japan having 17 hours as typical loads. At the advanced level, two-thirds of the countries had lighter teaching loads than the United States. The range was from a high of 32 periods weekly in Scotland to a low of just over 16 periods per week in Thailand and Japan. The typical assignment for U.S. mathematics teachers was 25 periods per week (five classes per day).

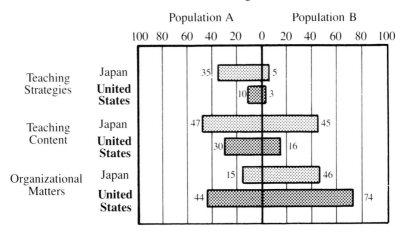

Percentage of Schools

| | | Population A | | | | | Population B | | | |

Activities at mathematics teachers meetings in Japan and in the U.S. are shown here. At both grade levels, the content of the meetings in the U.S. tended to focus on organizational matters while in Japan much more frequent attention was reportedly paid to consideration of teaching strategies and the subject matter to be taught.

Further indicators of differences in the professionalization of the working environments and in the status granted to teachers were provided by data on the in-school meetings of teachers and on what took place at those meetings. The number of meetings held in Japan was somewhat greater than in the United States and the content of those meetings differed markedly. School and departmental meetings of U.S. mathematics teachers most often dealt with managerial, administrative and school policy matters. Meetings of Japanese teachers tended to focus on the mathematics being taught and on specific instructional problems.

These data suggest that Japanese teachers were oriented more toward the subject matter and how to teach it and less toward the structure of the working situation. The U.S. mathematics teacher appeared to be more often isolated, at least formally, in dealing with instructional problems.

In a supplemental study, mathematics classes in the U.S. that were particularly effective were identified and their characteristics examined. "Effective" classes were those that achieved better than would be expected statistically, given such information as their scores at the beginning of the school year. It was found, for example, that at the eighth grade level, effective classes were more likely to be in schools having a departmental structure (that is, the schools were organized into subject matter departments, such as mathematics, science and English). In these "effective" classes, there was also evidence that their teachers followed the mathematics syllabus more closely and used more appropriate instructional strategies. Furthermore, teachers of these "effective" classes dealt more with professional resources and (perhaps most importantly) had lighter teaching loads. Such teachers were provided, presumably, the opportunity for the reflection and planning needed to organize their instructional programs.

At the twelfth grade level, the effective classes were taught by those well experienced in teaching college-preparatory mathematics. As one index of this experience, it was found that such teachers used more multiple methods of problem-solving than those in less effective classes. Furthermore, these effective teachers allocated more time to teaching *new* material, discussed difficulties arising from previously taught subject matter and provided opportunities for supervised study.

Minutes Per Week
(Per Class)

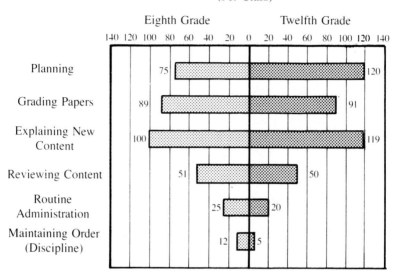

Eighth Grade | Twelfth Grade

140 120 100 80 60 40 20 0 20 40 60 80 100 120 140

Planning — 75 / 120

Grading Papers — 89 / 91

Explaining New Content — 100 / 119

Reviewing Content — 51 / 50

Routine Administration — 25 / 20

Maintaining Order (Discipline) — 12 / 5

How teachers' time was spent (in minutes per week for the *one* class sampled) shows a week dominated by preparation, grading papers and presenting mathematical content to the class, with less time spent on administrative and classroom management tasks. Multiplication of these time demands by five or six (the number of classes typically taught) to represent a typical U.S. teacher's load reveals the school to be a demanding workplace.

Deceptive Explanation Number Five: Quality of Mathematics Teaching

A final possible "explanation" of differences in student achievement focuses on the quality of the mathematics teaching that U.S. students received — the choices of instructional strategies, the resources utilized, and the specific activities that occupied the time of teachers and students.

In the Study, teachers estimated the time they devoted to various classroom activities. For the eighth grade U.S. teachers there was substantial variation both in total time commitment and in the way that time was used. The typical teacher spent from one to two hours per week outside of class in planning and another one to two hours in grading student papers for that class. Routine administrative duties such as taking attendance, making announcements, and setting up equipment required less than 30 minutes per week. Maintaining class order and disciplining students was reported to consume only about five minutes or less of an average class period. The teacher typically spent one to two hours of in-class time each week explaining new content and about half as much time reviewing old material.

At the advanced level in the U.S., teachers of Precalculus classes spent about two hours per week in preparation, one and a half hours in grading papers, two hours in class explaining new content, one hour in class reviewing old content and about 25 minutes per week in dealing with classroom administrative and management details. The profile was much the same for teachers of Calculus classes except that they typically spent an extra half-hour preparing and marginally less time reviewing and in administrative and management tasks.

We must keep in mind that these time expenditures were for only *one* class in a teaching load that typically consisted of 5 such classes. Thus, in addition to 25 to 30 hours per week of in-class work, teachers at both levels typically spent an estimated 10 to 24 hours outside of class each week in preparation and paper grading. This was in addition to other demands, such as conferring with individual students outside of class, attending meetings, and carrying out other non-class-related school responsibilities. Typically, the school as a workplace was an intensive consumer of time and filled with draining demands. For most U.S. mathematics teachers the workweek consistently went far beyond 40 hours.

71

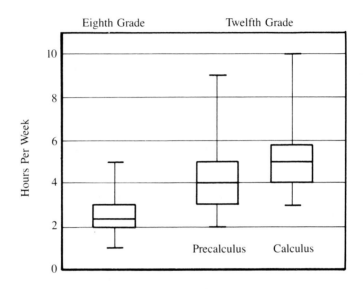

Eighth Grade   Twelfth Grade

Hours Per Week

Precalculus   Calculus

The amount of mathematics homework assigned in the United States is about two and one-half hours per week (30 minutes per day) at the eighth grade and almost twice as much (just over four hours per week) in the twelfth grade college preparatory mathematics classes. The data are presented here in "box-and-whisker" plots. The box includes the middle fifty percent of the classes reporting. So, at the eighth grade level, fifty percent of the mathematics teachers reported giving between two and three hours of homework per week. The line across the middle of each box represents the median or middle score. So, the median amount of mathematics homework assigned to the pre-calculus classes is close to four hours per week. That is, fifty percent of the classes were reported as having less than four hours of homework assigned. The lines protruding from each end of the box (the "whiskers") encompass the middle 90% of the classes reporting. So, we see that 90% of the eighth grade classes were assigned between one hour and five hours of mathematics homework per week. From an international point of view, the U.S. is about average in terms of the amount of mathematics homework assigned.

How students' time in class was spent (in minutes per week for the *one* class sampled) shows a week dominated by individual seatwork for the eighth grade U.S. students and by listening to lecture for twelfth grade students. Little time was spent on small group work at either grade level. ➡

Teachers were also asked to estimate the average time spent by students in selected in-class activities. For eighth grade U.S. students the majority of time was devoted to individual work at their seats or to listening to lectures and explanations by the teacher. A little less than one class period per week was typically spent on taking tests and quizzes. Very little time was spent in work with small groups. The pattern was similar for twelfth grade students in the U.S., although they spent somewhat less time on individual seat work and relatively more time listening to teacher explanations.

The majority of student in-class time was thus seen to be devoted to one of two types of activities: listening to teacher talk or individual work with minimal interaction with the teacher. The two types of activities often seem to have occupied distinct phases of the class period. Little place seems to have been given to guided, active discovery learning, in which students generated high-level questions and in which there was more of a balance between teacher and student subject-related talk. This pattern was not that different from that of other countries although the proportion of time spent on individual, in-class seat work was somewhat higher in the U.S.

With respect to mathematics homework, most eighth grade students reported spending two to three hours per week on homework for the one class sampled, while twelfth grade students spent from three to six hours per week for their one mathematics class. This amount of homework placed the U.S. in the middle range of the participating countries.

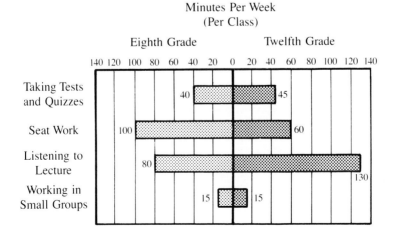

Minutes Per Week
(Per Class)

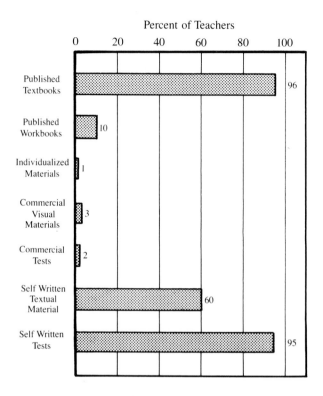

Percent of Teachers

The mathematics textbook in twelfth grade classes in the U.S. stands out as the most commonly and consistently used resource for instruction. Other commonly used resources are self-written tests and textual materials. Little use of commercial visual materials, commercial tests or materials for individualizing instruction was reported.

The student textbook is the predominant instructional resource in U.S. eighth grade mathematics classrooms, and is so reported by 90 percent of the teachers. Other material, such as workbooks, and worksheets, were reported to be of secondary importance. Other materials, such as films and laboratory materials, seldom were cited as a primary resource, and were rarely or never used by the vast majority of teachers. This was true even for topics such as geometry and measurement where such materials might be considered most helpful. ➡️

Mathematics instruction in U.S. classrooms is clearly textbook-driven. The textbook largely determines what is taught as well as what strategies are used in teaching it. Even when not actually determining the content of instruction, the textbook almost always defines the boundaries for instruction. According to the reasons given by teachers for using various instructional strategies and representations of mathematical content, inclusion in the textbook did not always guarantee that a strategy or content representation would be used in the classroom. However, exclusion from the textbook made it virtually certain that the strategy or representation would not be used. Mathematics instruction in the U.S. seemed to have been more tied to the textbook than was instruction in most other countries. One could speculate as to reasons for this. For example, it could be because of the absence in many parts of the U.S. of syllabi and examinations that are beyond the local school or district level.

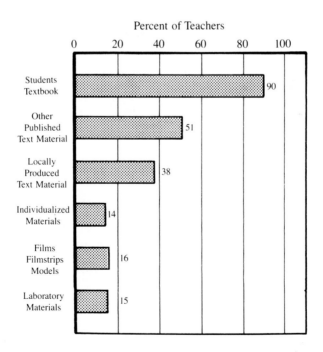

Percent of Teachers

Percent of Mathematics Classes

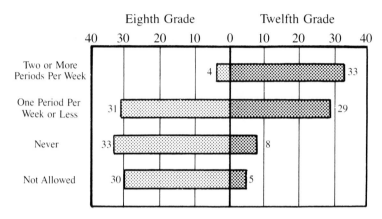

Eighth Grade | Twelfth Grade

	Eighth Grade	Twelfth Grade
Two or More Periods Per Week	4	33
One Period Per Week or Less	31	29
Never	33	8
Not Allowed	30	5

(Grade Twelve - about 20% of teachers did not respond)

The extent of calculator use in U.S. mathematics classrooms was found to be very limited at the eighth grade level, with very few classes reporting use more than one period per week and one-third of the classrooms reporting no use at all (data were collected in the school year 1981-1982). School policy prohibited the use of calculators in 30% of the eighth grade classrooms studied. Calculator use was much more frequent in U.S. advanced mathematics classrooms, as it was in other countries at this level.

Percent of Classes Encouraged to Use Calculators

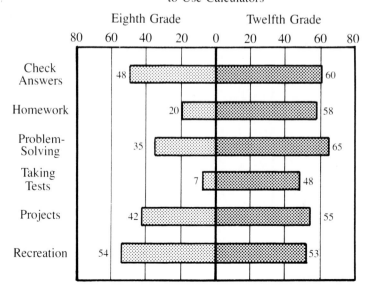

Eighth Grade | Twelfth Grade

	Eighth Grade	Twelfth Grade
Check Answers	48	60
Homework	20	58
Problem-Solving	35	65
Taking Tests	7	48
Projects	42	55
Recreation	54	53

Calculators and computers increasingly are becoming a part of everyday life. But data from the Study indicate that this technology had not entered the mathematics classroom in significant ways (at least when the data were collected in 1981-82). Only four percent of the eighth grade teachers reported using a calculator for two or more periods per week. About two-thirds of the eighth grade teachers reported calculators were never used or were not allowed in mathematics class. The most common uses of calculators were for recreational activities, checking exercises, and for projects.

The use of calculators was more prevalent at the advanced mathematics level. About one-third of the classes reported using them two or more periods per week and another one-fourth reported use averaging one period per week or less. Still, about 30 percent of the teachers reported that calculators were never used or not allowed in their classes. In twelfth grade, the most commonly reported uses of calculators in the classroom were for checking work, doing homework and solving mathematics problems in class, although about one-half of the twelfth graders were reported as being allowed to use calculators on examinations.

Three points should be noted about the limited role of calculators (and even more limited role for microcomputers) in U.S. mathematics classrooms. First, as was seen in the attitude data discussed earlier, this role for calculators may affect students' conceptions of the nature of mathematics and their plans for future participation in mathematics. Second, it seems clear that the use of calculators was not responsible for the low levels of achievement in U.S. classrooms since calculators were notably absent. Third, it seems that, at least at the time of the Study, calculators were not being used in those curricular areas where such use would be appropriate.

The kinds of calculator use in U.S. mathematics classrooms included checking answers and doing projects at the eighth grade level. Doing homework and problem-solving were more frequent uses reported at the advanced mathematics level. Recreational uses were rather common at both levels.

←

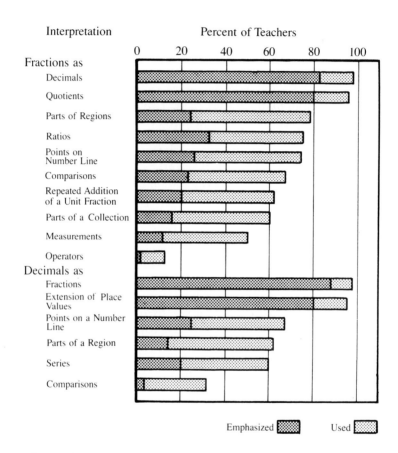

Interpretation — Percent of Teachers

Instruction on the concepts of common fractions and decimals in U.S. eighth grade classrooms was dominated by interpretations of the content that were more abstract and symbolic rather than concrete. Common fractions were often interpreted through references to decimals while decimals were often interpreted by reference to common fractions, hence one abstract concept was used to interpret another.

In addition to these more general aspects of instruction, detailed questionnaires were completed by teachers to provide precise information on strategies used to teach selected topics. While this information was very specific to the topics involved, some overall consistencies can be noted to further characterize the teaching of mathematics in U.S. schools.

At the eighth grade level, instruction appeared to be dominated by abstract and symbolic representations of content. In teaching most topics a variety of content representations were provided in the questionnaires, including some with strong perceptual elements and some with a more abstract or symbolic emphasis. For example, in teaching the concept of common fractions, fractions could be interpreted more perceptually (as parts of regions, points on a number line, measurements, etc.) or more abstractly (fractions as decimals, quotients, ratios, etc.). In almost all cases, the teachers more often chose the more abstract representations of content. Similar data were reported for teaching other topics — even those for which perceptual emphases are particularly appropriate, such as measurement and geometry.

Percent of Teachers

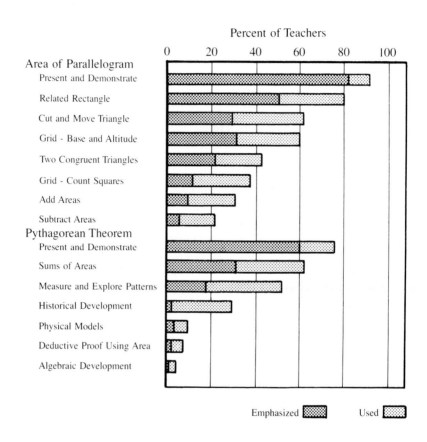

Area of Parallelogram
- Present and Demonstrate
- Related Rectangle
- Cut and Move Triangle
- Grid - Base and Altitude
- Two Congruent Triangles
- Grid - Count Squares
- Add Areas
- Subtract Areas

Pythagorean Theorem
- Present and Demonstrate
- Sums of Areas
- Measure and Explore Patterns
- Historical Development
- Physical Models
- Deductive Proof Using Area
- Algebraic Development

Emphasized   Used

Presenting and demonstrating the use of formulas was the most frequently emphasized method in instruction on selected formulas. More concrete strategies were neglected in favor of more abstract approaches.

Furthermore, procedures for various mathematical tasks were often developed by direct demonstration in a way that encouraged rote learning. Although active learning strategies such as constructing, measuring, counting and so on, were available for many topics, the single strategy most frequently emphasized by teachers was presenting and demonstrating procedures or stating definitions and properties — what has been characterized as "tell and show" approaches.

This use of abstract representations and of strategies geared to rote learning, along with class time devoted to listening to teacher explanations followed by individual seatwork and routine exercises, strongly suggests a view that learning for most students should be passive — teachers transmit knowledge to students who receive it and remember it mostly in the form in which it was transmitted. This might be considered as an approach of "rote teaching" and "rote learning." The attitude data discussed earlier support this perception of expected learning styles by showing that many students consider mathematics primarily a matter of rules and memorization. In the light of this, it is hardly surprising that the achievement test items on which U.S. students most often showed relatively greater growth were those most suited to performance of rote procedures (e.g., straight-forward computational items).

"The accomplishments of the students in twelfth grade who are enrolled in calculus or precalculus . . . are not in any sense an adequate or appropriate measure of how well the high school curriculum is delivering appropriate mathematics for college because there are so many students who aren't in twelfth grade mathematics but who nevertheless go on to college. They stop taking mathematics at the eleventh grade or earlier".

L. Steen
(1984)

The Study does raise some concerns about the quality of mathematics instruction in U.S. classrooms. Instructional approaches were dominated by the text. Little use was made of other resources, even calculators and microcomputers. A limited repertoire of instructional strategies was emphasized and those that were used often focused on the abstract and the rote.

It should also be noted, however, that information from the Study on how teachers go about their instruction on a day-to-day basis does not agree with their reported beliefs about what are the important goals of mathematics education. For example, at both grade levels, the majority of teachers reported that their most important objective was to assist students in developing a systematic approach to solving problems — an objective that has been promoted by such professional groups as the National Council of Teachers of Mathematics as a priority for the 1980s. It is reasonable to conclude that there are many factors at work, including the constraining effects of demanding workloads and the lack of professional support within school systems, that mitigate against teachers exhibiting the best practice that they are currently prepared to offer.

Surely these factors of instructional quality are strong candidates for explaining limited mathematics achievement. However, the pattern of factors seen here is similar to the patterns found elsewhere. Other countries, too, display strategies that emphasize the abstract, the rote, and reception learning. If these factors are to be seen as explanatory, they must be accompanied by other differences that make their impact more pronounced in U.S. classroom settings.

VI. THE UNDERACHIEVING CURRICULUM: CURRICULUM AS DISTRIBUTOR OF GOALS AND CONTENT

It would be very useful for decision makers to have available a set of standards for curriculum, with indicators of quality, that could be used by local schools and districts as yardsticks by which they could compare and measure their programs.

S. Hill
(1985a)

In the various critiques of U.S. education, the curriculum has received relatively little attention. Yet we believe that data from the Study provide convincing evidence that the mathematics curriculum deserves careful scrutiny and attention during this time of concern for educational renewal. In its goals, in its strategies and in its expectations for students, the Study has shown the U.S. school mathematics curriculum to be underachieving.

The functions of curriculum are many, as are its forms. The curriculum is not only what is intended to be taught, as reflected in syllabi and textbooks, it is also what is actually taught (implemented) in classrooms and what is attained by students as a result of that instruction. One of the main functions of curriculum, as intended and as implemented, is to distribute the content of the curriculum throughout the days and years of schooling according to a coherent and reasoned set of goals.

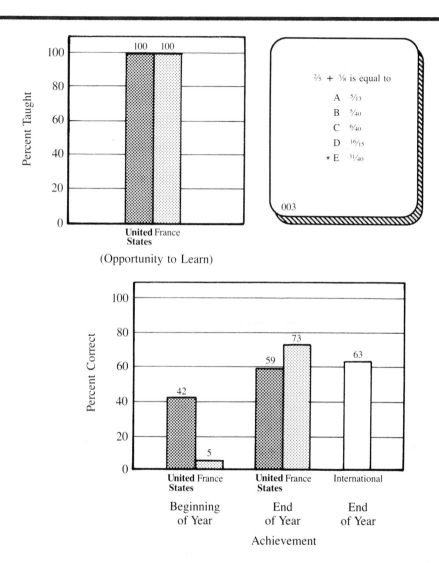

(Opportunity to Learn)

⅔ + ⅜ is equal to

A 5/13
B 5/40
C 6/40
D 16/15
★ E 31/40

003

Achievement

Curricular Intensity

The U.S. curriculum was found to have a different structure from other countries. At the lower secondary school level, Japan has a curriculum that emphasizes algebra. France and Belgium have curricula dominated by geometry and by fractions. By contrast, the U.S. allocates its time much more equally among the various topics. In no case does any one topic stand out as receiving markedly more time. At the upper secondary school level, Japan has a clear emphasis on calculus. Again, the U.S. has no one clear emphasis, but rather shares time among a variety of topics. At both levels, the U.S. curriculum may be characterized as a "low intensity curriculum" in terms of the time allocated to various topics.

Addition of common fractions was reported to be taught to all Population A students in France and the United States by the end of the school year. However, very different patterns of achievement were found in the two countries. In the United States, about 40% of eighth grade students were able to add these fractions at the beginning of the year (this topic is taught in the elementary grades in the U.S. and repeatedly revisited during junior high school). By the end of eighth grade, 59% of U.S. eighth graders were able to add these fractions. By contrast, very few French students could do this question at the beginning of the school year. At the end of the year, however, students in France achieved well — with a score of 73%. In France, common fractions receive little attention until the Population A year, where they are given intense coverage. In the earlier grades, emphasis is placed instead on decimal fractions — where instruction may be effectively combined with measurement activities involving the metric system.

←

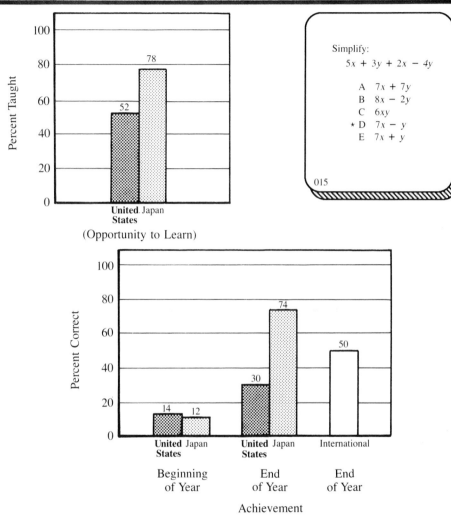

(Opportunity to Learn)

Achievement

Simplifying algebraic expressions (as represented by this item), was taught to 78% of Japanese seventh graders and to 52% of U.S. eighth graders. Japanese achievement rose by 62 points during the year, while U.S. achievement rose by only 16 points. The intense treatment given to algebra in Japan during the seventh grade is reflected in their achievement scores.

Subtracting integers (signed numbers) was reported to be taught to most U.S. classes by the end of eighth grade and to all Japanese classes by the end of seventh grade. In both countries, beginning of the year performance on this item was low. By the end of the year, U.S. achievement had increased by only 19 points while Japanese scores increased by 53 points! ➝

This same lack of intensity seen at the "macro" level of annual allocations of time was also seen at the "micro" level of daily and weekly allocations of time. Again, the pattern found was one of low intensity or emphasis on individual topics as reflected in the allocation of a few periods, often only one, to each of many topics rather than the allocation of many periods to a few topics. Instead of being characterized by "single concept lessons" (lessons focusing on only one major concept at a time), the U.S. curriculum appears to feature "single lesson concepts" (concepts and topics so fragmented that they can be taught in one or two lessons so that the next can be spent on a different topic). These patterns in U.S. mathematics curricula are not typically seen in the curricula of other countries, especially of high-achieving countries, such as Japan.

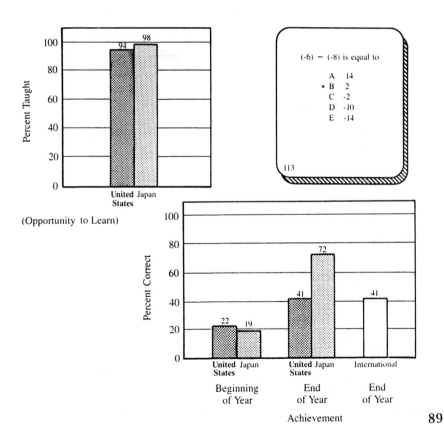

(Opportunity to Learn)

Achievement

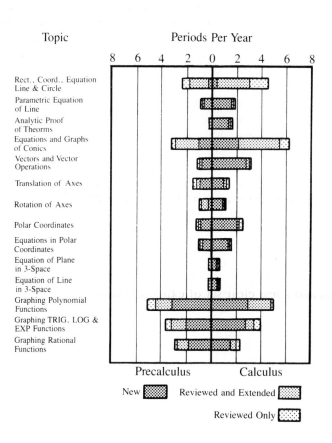

Topic   Periods Per Year

8   6   4   2   0   2   4   6   8

Rect., Coord.. Equation Line & Circle

Parametric Equation of Line

Analytic Proof of Theorms

Equations and Graphs of Conics

Vectors and Vector Operations

Translation of Axes

Rotation of Axes

Polar Coordinates

Equations in Polar Coordinates

Equation of Plane in 3-Space

Equation of Line in 3-Space

Graphing Polynomial Functions

Graphing TRIG. LOG & EXP Functions

Graphing Rational Functions

Precalculus   Calculus

New ▨   Reviewed and Extended ▨

Reviewed Only ▨

A lack of intensive treatment of topics in analytic geometry was the norm in U.S. Precalculus classes (where a full year of calculus was not being studied). A pattern of somewhat more intensive, extended coverage of topics was seen in the Calculus classrooms.

As the data reveal, Japanese students showed remarkable achievement gains during the school year — gains unmatched by U.S. students. France showed a similar pattern of emphasis, with clear goals and high growth for common fractions. (See, for example, the figures *Addition of common fractions* and *Simplifying algebraic expressions.*)

At the advanced mathematics level, the U.S. sample included classes studying an entire year of calculus as well as those that covered a variety of precalculus topics. The Calculus classes followed the syllabus for one of the AP (College Entrance Examination Board Advanced Placement) calculus examinations. These classes, therefore, shared a common syllabus that emphasized a specified set of content, clear goals to be met in a definite period of time, and clear expectations for student achievement. Furthermore, such achievement was to be assessed by an examination beyond the control of local teachers and schools. The Precalculus classes operated in a different context, dealing with a more varied, locally determined set of topics, a more diffuse set of goals and a less clear set of expectations for student attainment.

*The [school mathematics]
curriculum is chaotic and its
implementation one of an-
archy .*

S. Hill
(1985b)

*The problem of curriculum is
to economize scarce learning
potential by making the most
judicious and appropriate selection of study content. Human intelligence is too rare
and precious a thing to squan-
der on a haphazard program
of instruction.*

Philip H. Phenix
(1958)

The Diffusion of Goals and Expectations

Along with the lack of intensity in the distribution of content there appears to be a diffusion of goals and expectations for that content. It seems to have been the case that, just as the content of any reasonably large segment of instruction was mixed, there often were no clear goals for that segment of instruction. Even when definite goals were present, there appeared to be no clear expectation that those goals must be met during that particular segment of instruction. There seems to have been the sense that if goals were not met and content not mastered during the current year, there would be chances to do so in subsequent years.

Such diffusion was less often found in other countries. In Japan, significant amounts of algebra were taught for the first time during the Population A year (seventh grade in Japan). That year was the one time at which the goals for beginning algebra content were to be met and there were clear expectations for what the Japanese students must accomplish.

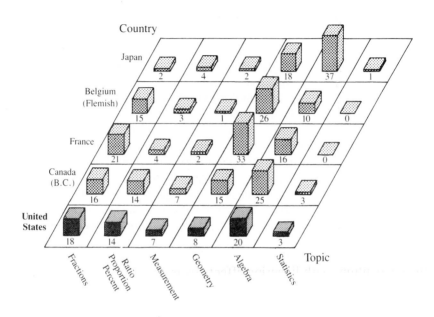

Country

Japan

Belgium
(Flemish)

France

Canada
(B.C.)

United
States

Topic

Fractions   Ratio   Measurement   Geometry   Algebra   Statistics
           Proportion
           Percent

	Fractions	Ratio Proportion Percent	Measurement	Geometry	Algebra	Statistics
Japan	2	4	2	18	37	1
Belgium (Flemish)	15	3	1	26	10	0
France	21	4	2	33	16	0
Canada (B.C.)	16	14	7	15	25	3
United States	18	14	7	8	20	3

   Intensity of mathematics instruction at the Population A level (eighth grade in the U.S.), based on teacher reports of anticipated percentage of class periods devoted to various topics, shows distinct patterns across countries. The Japanese curriculum provides their seventh graders with an intensive introduction to algebra. The Belgian and French programs focus on geometry and extensive work on common fractions (decimal fractions were dealt with in the elementary grades.) Correspondingly dramatic growth in achievement during the school year takes place for these topics in Japan, France and Belgium. By contrast, the U.S. curriculum has little intensity, with little sustained attention paid to any aspect of mathematics.

   Patterns of intensity of mathematics instruction at the Population B level (twelfth grade college preparatory mathematics in the United States), based on the anticipated number of class periods devoted to various topics, highlight a distinct focus on calculus in Japan, with considerable attention paid to the subject in Belgium and New Zealand, as well. British Columbia provides an intense approach to algebra, but does little with calculus. In the United States, coverage appears to be spread over a variety of topics, with little concentrated attention devoted to any (with the exception, of course, of the Advanced Placement Calculus Program, not reflected here).━━━━▶

The Lingering of Goals and Content

The figures *Intensity of mathematics instruction* and *Patterns of intensity* make clear one other characteristic of U.S. school mathematics curricula. In view of the time spent on fractions, ratio and proportion and percent, the U.S. seems to have a curriculum that is "driven" by arithmetic at the Population A level. By contrast, Japan has an "algebra-driven" curriculum. At the Population B level, the U.S. seems to have an "algebra-driven" curriculum while most other countries have a "calculus-driven" curriculum.

The U.S. eighth grade curriculum, therefore, is much more like a curriculum of the last years of elementary school while that of Japan, and many other countries, resembles that of the first years of secondary school. At the twelfth grade level, the U.S. curriculum is much more like that of early years of secondary school while the curriculum of most other countries is more like that of beginning college level.

In the lower secondary school, arithmetic continues to drive the curriculum, with lingering effects from the elementary school years. As one mathematics educator has observed, "The propensity of U.S. schools to prolong the teaching of arithmetic beyond the elementary grades reflects our evidently compelling need to continue instruction on the subject until mastery is obtained."

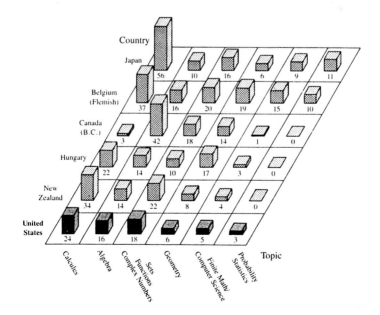

*"Experience . . . points to
the fact that our schools
may be wasting precious
years by postponing the
teaching of many impor-
tant subjects on the ground
that they are too difficult . . .
[T]he basic ideas that lie
at the heart of all science
and mathematics . . . are as
simple as they are powerful.
To be in command of these
basic ideas, to use them effectively, requires a continual deepening of one's
understanding of them that
comes from learning to use
them in progressively more
complex forms . . . . A curriculum as it develops
should revisit these basic
ideas repeatedly, building
upon them until the student
has grasped the full formal
apparatus that goes with
them . . . There is much still
to be learned about the
'spiral curriculum' that turns
back on itself at higher levels,
and many questions still to
be answered . . ."*

J. Bruner
(1977)

The Spiral Curriculum

The pattern of U.S. mathematics curricula returning again and again to the same collection of unmastered skills and concepts might be defended as an implementation of the logic of the spiral curriculum. The spiral curriculum structure, as set forth in the classic statement of Jerome Bruner quoted earlier here, assumes: (1) simple but powerful basic ideas central to the curriculum, (2) a "continual deepening" of the understanding of these basic ideas, (3) learning to apply the ideas "in progressively more complex forms", (4) the repeated "revisiting" of the basic ideas, and (5) the ultimate grasping by the student of the "full formal apparatus" that goes with the basic ideas. This logic is simple, elegant and intellectually appealing.

If, in fact, designers of school mathematics curricula have consciously tried to implement a spiral structure, the theory has encountered the realities of schooling and the market forces that shape the development of mathematics textbooks. In the process, it has become something other than elegant and, finally, something other than appealing. The "simple but powerful basic ideas" have become fragments of content, often computationally-oriented, that might be assembled incrementally into larger structures. However, the resulting content has often had little power or importance and only the simplicity produced by fragmentation. The application of basic ideas in "progressively more complex forms" has become the assembling of fragments in the incremental attainment of more complex computational skills.

"I didn't learn much this year that I didn't already know from last year. Math is my favorite class, but we just did a lot of the same stuff we did last year."

Andy, 4th Grade
(1986)

My own conclusion, as a long-term student of curriculum reform, is that there has not been intensive, sustained attention to the content of elementary and secondary education for some time.

J. I. Goodlad
(1984b)

The "continual deepening of understanding" seems to have been sacrificed to the demands for content suitable for rote reception learning. Grasping for the "full formal apparatus" of ideas appears to have been transformed into the rote manipulation of abstract formalisms deemed suitable for the algorithms of computational skills. All that has remained of the original logic is the continual revisiting of the same ideas. The logic of the spiral curriculum has degenerated into a spiral of almost constant radius — "a curriculum that goes around in circles."

It was the continual revisiting of ideas that seemed to legitimate the lingering of goals and content. It was the transformation of the continual deepening of basic ideas that served as a rationale for the diffusion of goals and expectations. It was the motif of the spiral that validated the distribution of mathematical content over the grades when a more intensive organization of the content might have accomplished a higher level and more lasting mastery. Perhaps the greatest irony is that a curricular construct conceived to *prevent* the "postponing of teaching many important subjects on the grounds that they are too difficult" has resulted in a treatment of mathematics that has postponed, often indefinitely, the attainment of much substantive content at all.

The misapplied philosophy of the spiral curriculum has had its impact not only on school mathematics curricula but on the undergraduate curricula of colleges as well. When the curricula of colleges are viewed as just one more arm of the spiral, as one further chance for the delayed attainment of diffuse goals, then the continued presence of high school remedial courses in U.S. collegiate mathematics programs is assured.

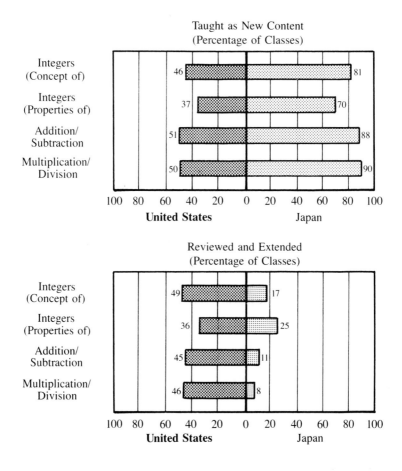

Taught as New Content
(Percentage of Classes)

Integers (Concept of): United States 46, Japan 81
Integers (Properties of): United States 37, Japan 70
Addition/Subtraction: United States 51, Japan 88
Multiplication/Division: United States 50, Japan 90

100 80 60 40 20 0 20 40 60 80 100
United States Japan

Reviewed and Extended
(Percentage of Classes)

Integers (Concept of): United States 49, Japan 17
Integers (Properties of): United States 36, Japan 25
Addition/Subtraction: United States 45, Japan 11
Multiplication/Division: United States 46, Japan 8

100 80 60 40 20 0 20 40 60 80 100
United States Japan

The teaching of integers (signed numbers) takes place for the first time in the seventh grade in Japan, and impressive amounts of learning of the subject occur. By contrast, in the United States, algebra topics are typically reviewed and extended during eighth grade, with relatively little learning taking place.

Alternative Curricular Organization

The organization of the mathematics curriculum found elsewhere (for example, France and Japan), suggests alternative structures to consider. For example, a more "blocked" or "sequential" organization would allocate significant amounts of content to specified places in the mathematics program and facilitate intense coverage of that subject matter. The allocation of content to such blocks also promotes establishing clear expectations for achievement prior to moving on to the next substantial body of mathematics.

The imagery called for here, as an alternative to spiralling, may be that of the *concentric curriculum* — with "concentric" suggesting arrays of circles with the same center but of increasing diameters. That is to say, "single but powerful basic ideas," such as that of a variable, or of a function, are imbedded in layers of the curriculum that are ever-expanding and deepening. This imagery emphasizes that concepts are not simply re-visited and then set aside until the next visit. Rather, each encounter with the concept significantly moves learning to a "continual deepening" of understanding.

A variation of the imagery invoked by a concentric structure may be that provided by a rock thrown into a lake. The resulting pattern of waves emanating from the point of impact suggests the ever expanding and broadening development of key concepts throughout a well articulated and effectively implemented curriculum from the earliest grades through graduate school.

VI. THE UNDERACHIEVING CURRICULUM: CURRICULUM AS DISTRIBUTOR OF OPPORTUNITY-TO-LEARN

. . . the central problem for today and tomorrow is no longer access to school. It is access to knowledge for all. The true challenge is that of assuring both equity and quality in school programs.

J. Goodlad
(1985b)

I argue . . . not for a homogeneous curriculum, nor for indifference to the special problems that burden some children's learning, but for a sense of commitment to intellectual development as a proper goal for all children, not just the gifted.

D. Ravitch
(1985)

Curriculum, taken in the larger sense of including policies that determine types of classes as well as what will be studied in each type of class, serves the social function of distributing opportunities to learn to students. The curriculum adopted for a given class sets boundaries on the attainments of individual students by determining the content to which they will be exposed and therefore the learning opportunities afforded them.

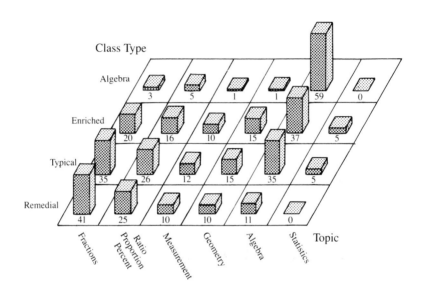

Class Type

Algebra

3   5   1   1   59   0

Enriched

20   16   10   15   37   5

Typical

35   26   12   15   35   5

Remedial

41   25   10   10   11   0

Fractions   Ratio Proportion Percent   Measurement   Geometry   Algebra   Statistics   Topic

Four dramatically different eighth grade mathematics programs are provided students in the United States, as this figure suggests. Students in Remedial classes are offered content that is predominantly arithmetic. The Typical eighth grade program has a slightly greater exposure to geometry and considerably more coverage of algebra than offered to remedial classes. Enriched classes receive still more algebra and less exposure to arithmetic. The Algebra classes are provided a great deal of instruction in algebra — usually a standard high school freshman algebra course. Statistics receives scant attention in any of the four programs.

In U.S. eighth grade mathematics, four distinct types of curricula were identified based on teacher data and analyses of the textbooks used — Remedial, Typical, Enriched and Algebra classes. Each of these curricula allocates differing amounts of time to differing mixes of topics, resulting in instructional programs with vastly different mathematical content. (See *Four dramatically different eighth grade mathematics programs. . . .*) For example, the program for Remedial classes is dominated by the arithmetic of grade school, such as common and decimal fractions and percentage. Little instructional time is devoted to topics in geometry or algebra. Typical classes still receive instruction on arithmetic, but some attention is paid to geometry and algebra as well. In the Enriched classes less time is allocated to arithmetic than in the prior programs, and slightly more time is given to algebra. The Algebra classes are provided with an introduction to high school algebra. However, such students are at the same time deprived of instruction in geometry — these students will receive virtually no instruction on this topic until their high school sophomore geometry course. It is also important to note that the topic of statistics receives virtually no attention in the Algebra classes, and little time is given to this topic in the other classes either.

Percent of Items

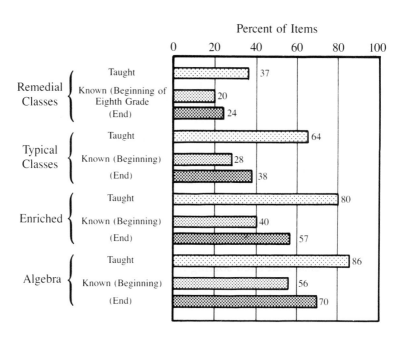

Four different class types are found in eighth grade mathematics in the United States, each with a different mathematics curriculum. In this graph, the amount of algebra taught and learned by students in each class type is shown. Students in the Remedial classes were taught only about one-third of the algebra on the international test, and very little algebra was learned during the year. On the other hand, students in the Algebra classes were taught almost all of the algebra on the test, and a great deal was learned. More extensive tracking was found at this level in the United States than in any other country in the Study.

The relationship between opportunity-to-learn and achievement that was found in the eighth grade U.S. classrooms is demonstrated by the data for algebra that are presented in the figure *Four different class types.* . . . The large differences between class types in the amount of algebra taught is reflected in the correspondingly large differences in how much algebra was learned. Indeed, by the *end* of the school year, each class type achieved about what the next more advanced group had achieved at the *beginning* of that year. That is, each successively more advanced curriculum type appeared to produce student achievement that was about one year ahead of the achievement by the preceding class type.

The dramatic differentiation of opportunity and achievement that is found in U.S. mathematics classes is disturbing. However, some may view such differentiation as essential to the efficient implementation of instruction, enhancing achievement by creating relatively homogeneous classes for which appropriate mixes of content can be provided so that each student is provided with the most appropriate challenge.

Was it the case that such sorting into differing curricular types had only the effect of increasing instructional efficiency in mathematics? And was it the case that this sorting served to make available to each student the mathematics learning tasks that provided the most appropriate challenge?

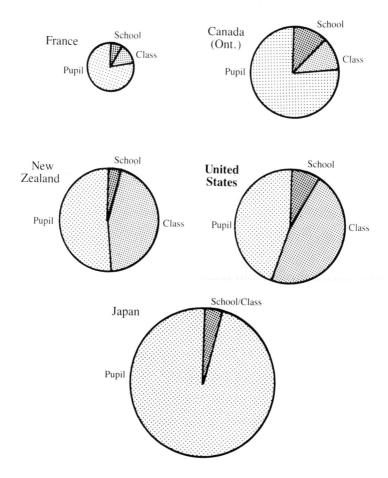

These circles depict for each country the variance components associated with the core pretest. (For most countries, this test consisted of 40 items. For Japan, it had 60 items.) The areas of the circles are roughly proportional to the total amount of score variation. The slices represent the amount of total variation that is attributable to differences between students, classrooms and schools. Since these are pretest data, the variance components represent how students were allocated to schools and classrooms and not to differences in teaching during the school year. Very similar patterns were found in the posttest data, as well.

One answer to the first of these questions is provided by the "pies" shown in the figure *These circles depict for each country. . . .* When the United States is contrasted with other countries, several points may be noted. First, a relatively low proportion of the variance in U.S. achievement is attributable to differences between schools. That is, compared to those of other countries, U.S. schools appear to differ relatively little in the mathematics achievement of their students.

However, vast differences in achievement were found from class to class *within* U.S. schools. The comparatively large slice of the variance "pie" that is attributable to differences between classrooms within schools supports the data presented earlier on the differences in opportunity-to-learn created by the four eighth grade mathematics class types.

Perhaps most disturbing, the slice of the "pie" representing differences between students was less in the U.S. than in any other country. It appears that curricular differentiation between schools, and especially between classes within schools, has set boundaries on what each individual can achieve by his or her own efforts. Apparently, there was too little of the "pie" left to appropriately reward individual effort after the limits were set by curricular differentiation.

In short, it was *not* the case that differences in opportunity-to-learn mathematics in the sampled classrooms of the Study resulted in increased efficiency and instructional effectiveness. These differences resulted instead in a deceptive appearance of equality by allocating substantial differences of opportunity between classes within schools.

Was it the case that this sorting and differentiation of opportunity at least served to provide each student with the most appropriate challenge in the opportunity for mathematics learning? The answer again is that it was *not* likely that the sorting that took place was efficacious in matching ability to opportunity.

Where do American schools stand with respect to curriculum and tracking? We have compromised. Our rhetoric, our commitment to the comprehensive school, and our provision in most school systems of an undifferentiated high school diploma all suggest a decision against tracking. However, in reality we have considerable tracking in our high schools, even if it is not always formally so labeled. Comprehensive high schools usually house several quite different sets of courses in which expectations and standards vary considerably.

D. P. Resnick and L. B. Resnick
(1985)

One attempt to assess the efficiency of the sorting was to look at the prior achievement of students in the four class types in the U.S. This was done by analyzing the patterns of achievement scores in arithmetic at the beginning of the school year for students in each class type. In general, it was found, as expected, that students in the more advanced class types had higher arithmetic scores. However, considerable overlap in these scores among the class types was also noted. For example, some students in Remedial classes had higher arithmetic scores than some students in Typical classes; some students in Enriched classes had higher scores than those in Algebra classes, and so on. It was also found that the lowest scoring students in these Algebra classes had lower arithmetic scores than virtually *all* the students in the Typical classes and 75% of the students in the Remedial classes.

Therefore, not only did the distribution of opportunity-to-learn provided by the U.S. eighth grade mathematics curricula serve to limit the efficacy of individual ability, but the sorting was not efficiently done. More able students were not always placed into more demanding and opportunity-laden classes. These data raise serious questions about the merits of curricular differentiation, especially in grades as early as junior high school. Under current policies, large proportions of U.S. students are being deprived of opportunities to participate in the best mathematics programs that schools have to offer.

VIII. THE ROAD TO RENEWAL: ON BECOMING A NATION OF MATHEMATICS ACHIEVERS

> *If getting better schools means doing better what we now do, the job is relatively easy. But if it means living up to our rhetoric about what education should be, then the challenge is one of the greatest this nation has faced.*
>
> J. Goodlad
> *(1986)*

The case for change has been set out from the data of the Second Mathematics Study. A number of propositions are now presented. Some follow clearly from the Study's data. Others seem likely to be true but are more dependent on the judgment of the authors of this report:

1. *A restructuring of the mathematics curriculum appears to be called for.* The effectiveness of the spiral curriculum as implemented in U.S. classrooms needs to be reconsidered and alternative arrangements explored. Any new organization of curriculum should increase the intensity of content coverage and establish clear expectations for student attainment. Definite accomplishment milestones and leaving points need to be established in order to minimize the lingering of unmastered content from year to year.

2. *The content of the mathematics curriculum needs to be re-examined and revitalized.* The domination of the lower secondary school curriculum by the arithmetic of the elementary school has resulted in a program that, from an international point of view, is very lean. The curriculum should be broadened and enriched by including a substantial treatment of topics such as geometry, probability, statistics and algebra, as well as promoting higher-level process goals such as estimation and problem-solving.

3. *Clear standards for achievement must be established at each grade level in order to create an institutionalized climate of expectation to which students will respond.* The effective implementation of these standards likely will require new approaches to educational assessment.

4. *The practice of early sorting of students into curricular tracks that lead to vastly different opportunities to learn high school mathematics must be carefully re-examined.* Currently, significant proportions of young students are being assigned to mathematics classes that offer relatively little opportunity to learn the mathematics content needed for success in high school and beyond.

. . . a need implied in these (SIMS) data is for clearly defined and widely accepted goals, a framework upon which to hang a curriculum, which, while it allows for local adaptation, is based upon a set of broadly agreed upon priorities that teachers can use as guides.

S. Hill
(1985c)

In the light of the facts and implications of the SIMS data, it is time for policy makers to examine, free of the biases of turf, politics and self-interest, the sacred cow of diversity and local choice cited as justification for the autonomy in curricular objectives accorded local school jurisdictions. While authority and legal control belongs at state and local levels, educational policy is inescapably related to national economic concerns and national security.

S. Hill
(1985d)

5. *The role of textbooks as curriculum guides ought to receive renewed attention.* In any consideration of curriculum reform, textbook publishers are critically important agents.

6. *The factors leading to low enrollments in advanced mathematics courses must be analyzed.* This would include a careful study of student beliefs and attitudes, of what shapes their conception of mathematics and what determines their commitment to its study. Data from other surveys (notably the National Assessment of Educational Progress) indicate that children begin school with a keen interest in mathematics, but this interest declines significantly through the grades. Ways must be found to help children retain this excitement and nourish it in order that they remain engaged in the study of mathematics.

7. *Status and rewards must be granted to teachers commensurate with the demands of their profession.* A significant part of such efforts should include the reduction of the professional isolation that was shown by the Study to characterize the daily life of many U.S. mathematics teachers.

8. *Ways need to be found to upgrade the school classroom as a workplace.* Data from the Study strongly suggest that U.S. mathematics teachers are being prevented, by such factors as heavy teaching loads (very heavy by international standards), from carrying out the best teaching that they are prepared to provide.

9. *Professional development programs for mathematics teachers must be improved.* Such programs would include ways to broaden the repertoire of teaching strategies that promote mathematics learning as an active rather than a passive enterprise.

10. *Technology, including calculators and computers, should find an appropriate place in the mathematics classroom.* The use of

A curriculum is more for teachers than it is for pupils. If it cannot change, move, perturb, inform teachers, it will have no effect on those whom they teach.

J. S. Bruner
(1977a)

such technology will promote the upgrading of the mathematical content of the curriculum as well as assist in the effective teaching of that content.

U.S. school mathematics, like U.S. schooling in general, is somewhat like Gulliver bound by the thousand cords of Lilliput. Complex and sustained efforts are needed to help break the many bonds that prevent us from providing our children and youth the kind of mathematics education required today. Much could be accomplished by a few sensible demands, by the expectation that students can accomplish far more than they now do, and by clear goals for a reorganized and revitalized mathematics curriculum.

It is time we took the road toward a renewal of school mathematics and started on the way to becoming a nation of mathematics achievers. It is time to restructure and revitalize the mathematics curriculum. It is time to find ways to make the workplace of teachers more productive and more rewarding. It is time to take a careful look at the long-term effects of the early sorting of students into mathematics programs of dramatically different content and quality. It is time to re-engage our young people in mathematics as a field of study that is critically important for their futures. It is time, and past time, for change.

*   *   *   *   *   *   *   *

Today we maintain ourselves. Tomorrow science will have moved forward yet one more step, and there will be no appeal from the judgement which will then be pronounced on the uneducated.

A. N. Whitehead
(1916)

Appendix I
SIMS Committees
National Mathematics Committee

James Fey, University of Maryland (Chairman)
Joe Crosswhite, Northern Arizona University
John A. Dossey, Illinois State University
Floyd Downs, Hillsdale High School, San Mateo, California
Edward Kifer, University of Kentucky
Curtis C. McKnight, University of Oklahoma (National Research Coordinator)
Jane Swafford, Northern Michigan University
Kenneth J. Travers, University of Illinois at Urbana-Champaign
A. I. Weinzweig, University of Illinois at Chicago
James Wilson, University of Georgia
Richard Wolf, Teachers College, Columbia University

National Technical Advisory Panel

Edward Kifer, University of Kentucky (Chairman)
Leigh Burstein, University of California—Los Angeles
Robert Linn, University of Illinois at Urbana-Champaign
William Schmidt, National Science Foundation (on leave from
    Michigan State University)
Jack Schwille, Michigan State University
Richard Wolf, Teachers College, Columbia University
Richard Wolfe, Ontario Institute for Studies in Education

Classroom Processes Working Groups

Eighth Grade: Thomas J. Cooney, University of Georgia (Chairman)
Nicholas Branca, San Diego State University
John A. Dossey, Illinois State University
James Hirstein, Slippery Rock University
Tom Kieren, University of Alberta
David Robitaille, University of British Columbia
Leslie Steffe, University of Georgia
Alba Thompson, Illinois State University
Paul Weichsel, University of Illinois at Urbana-Champaign

Twelfth Grade: John A. Dossey, Illinois State University (Chairman)
Peter Braunfeld, University of Illinois at Urbana-Champaign
L. Ray Carry, University of Texas at Austin
Thomas Cooney, University of Georgia
Douglas Grouws, University of Missouri, Columbia
John LeDuc, Eastern Illinois University
Norman Webb, University of Wisconsin at Madison

118

National Advisory Panel

Peter Braunfeld, University of Illinois at Urbana-Champaign
Robert Davis, University of Illinois at Urbana-Champaign
Ross Finney, Educational Development Center, Newton, Massachusetts
Douglas Grouws, University of Missouri, Columbia
Carl Guerriero, Dickinson College (formerly of Pennsylvania State Department of Education)
Shirley Hill, University of Missouri, Kansas City
George Immerzeel, University of Northern Iowa
Horacio Porta, University of Illinois at Urbana-Champaign
Alan Purves, State University of New York at Albany
Leslie Steffe, University of Georgia
Dorothy Strong, Chicago Board of Education
Bruce Vogeli, Teachers College, Columbia University
Paul Weichsel, University of Illinois at Urbana-Champaign
Russell Zwoyer, University of Illinois at Urbana-Champaign

U.S. National Coordinating Center Staff
(During time period 1976-1986)

Department of Secondary Education, College of Education
University of Illinois at Urbana-Champaign

Li-Chu Chang (Deceased)
Gullaya Dhompongsa
Larry Dornacker
Roberta L. Harrison
James E. Hecht
Dorothea Helms
James J. Hirstein
Chantanee Indrasuta
Gail Jaji
Del Jervis
Elizabeth Jockusch
Lynn Juhl

Leigh Little
Denise McCoy
M. David Miller
Claudia Nieto
Judy Ruzicka
Kazem Salimizadeh
Horace Smith
Sally Spaulding
Peter M. Staples
Nongnuch Wattanawaha
Ian Westbury
John Williams

Appendix II

The International Association for the Evaluation of Educational Achievement

The International Association for the Evaluation of Educational Achievement (IEA), is an international, non-profit-making scientific association incorporated in Belgium for the principal purposes of: (a) undertaking educational research on an international scale; (b) promoting research aimed at examining educational problems in order to provide facts that can help in the ultimate improvement of educational systems; and (c) providing the means whereby research centers in the various member countries of IEA can undertake co-operative projects. The current chairman of the IEA Council is Alan C. Purves, State University of New York, Albany, U.S.A.

The Mathematics Project Council, responsible for the Second Mathematics Study, was chaired by Roy W. Phillipps of the New Zealand Department of Education. Robert Garden, also of the New Zealand Department of Education, was International Project Co-ordinator for the Study. Kenneth J. Travers was Chairman of the International Mathematics Committee (IMC), which designed the Second International Mathematics Study and developed the international instruments. Other members of the IMC were: Sven Hilding, Sweden; Edward Kifer, United States; Gerard Pollock, Scotland; Tamas Varga, Hungary; and James Wilson, United States. A. I. Weinzweig, United States, was consulting mathematician and Richard Wolfe, Canada, was consulting psychometrician to the IMC.

Appendix III

Participating Countries

Country	National Research Coordinator	Council Member	Center
Australia*	Malcolm Rosier	John Keeves	Australian Council for Educational Research, Hawthorn, Victoria
Belgium (Flemish)	Christiana Brusselmans-Dehairs	A. De Block (1976-81) J. Heene	Seminaire en Laboratorium voor Didactiek, Gent
Belgium* (French)	George Henry	Gilbert De Landsheere	Université de Liège au Sart Tilman
Canada (B.C.)	David Robitaille	Joyce Matheson	Ministry of Education, Victoria
Canada (Ontario)	Leslie McLean	Bernard Shapiro	Ontario Institute for Studies in Education, Toronto
Chile	Maritza Jury Sellan	Marino Pizarro	University of Chile, Santiago
England*/Wales	Michael Cresswell (1978-83) Derek Foxman	Clare Burstall	National Foundation for Educational Research, Berkshire
Finland*	Erkki Kangasniemi	Kimmo Leimu	Institute for Educational Research, University of Jyväskylä, Jyväskylä
France*	Daniel Robin	Daniel Robin	Institut National de Récherche Pédagogique, Paris
Hong Kong	Patrick Griffin (1976-83)	M. A. Brimer	University of Hong Kong, Hong Kong
Hungary	Julia Szendrei	Zoltán Báthory	Országos Pedagógiai Intézet, Budapest
Ireland	Elizabeth Oldham	Elizabeth Oldham	Trinity College, Dublin

Country	National Research Coordinator	Council Member	Center
Israel*	Arieh Lewy (1976-83) David Nevo	Arieh Lewy (1976-83) David Nevo	Tel-Aviv University Tel-Aviv
Ivory Coast	Sango Djibril	Ignace Koffi	Service D'Evaluation, Abidjan
Japan*	Toshio Sawada	Hiroshi Kida	National Institute for Educational Research, Tokyo
Luxembourg	Robert Dieschbourg	Robert Dieschbourg	Institut Pédagogique, Walferdange
Netherlands*	Hans Pelgrum	Egbert Warries	Twente University of Technology, Enschede
New Zealand	Athol Binns	Roy W. Phillipps	Department of Education, Wellington
Nigeria	Wole Falayajo	E. A. Yoloye	University of Ibadan, Ibadan,
Scotland*	Gerard Pollock	W. Bryan Dockrell	Scottish Council for Research in Education, Edinburgh
Swaziland	Mats Eklund (1976-81) P. Simelane	Mats Eklund (1976-81) P. Simelane	William Pitcher College, Manzini
Sweden*	Robert Liljefors	Torsten Husén (1976-83) Inger Marklund	University of Stockholm, Stockholm
Thailand	Samrerng Boonruangrutana	Pote Sapianchai	Institute for the Promotion of Teaching Science and Technology, Bangkok
United States*	Curtis McKnight	Richard Wolf	Teachers College, Columbia University New York

*Also participated in First International Mathematics Study. The Federal Republic of Germany took part in the First Study only.

Appendix IV
United States Reports

†*United States Summary Report: Second International Mathematics Study.* 1985.

†*Technical Report I,* Item Level Achievement and OTL Data, May 1985

Technical Report IV, Instrument Book, Achievement Tests and Background Questionnaire, (December 1985)

Technical Report V, Instrument Book, Classroom Process Questionnaires (December 1985)

†*Detailed National Report,* (December 1986)

Monographs of the *Journal for Research in Mathematics Education,* National Council of Teachers of Mathematics (to appear, 1987)

Classroom Processes in School Mathematics:
Volume I: Eighth Grade
Volume II: Advanced Mathematics

(Note: The last two reports are funded by grants from the National Science Foundation.)

†Published by Stipes Publishing Company, 10 Chester Street, Champaign, IL 61820

Articles:

Travers, Kenneth J. and Curtis C. McKnight, "Mathematics Achievement in U.S. Schools: Preliminary Findings From the Second IEA Mathematics Study," *Phi Delta Kappan,* February 1985, pp. 407-413.

McKnight, Curtis C., Kenneth J. Travers, F. Joe Crosswhite, and Jane O. Swafford, "Eighth-Grade Mathematics in U.S. Schools: A Report From the Second International Mathematics Study," *Arithmetic Teacher,* Vol. 32, No. 8, April, 1985, pp. 20-26.

McKnight, Curtis C., Kenneth J. Travers, and John A. Dossey, "Twelfth-Grade Mathematics in U.S. High Schools: A Report From the Second International Mathematics Study," *Mathematics Teacher,* Vol. 78, No. 4, April, 1985, pp. 292-300 (cont. 270).

Travers, Kenneth J., Curtis C. McKnight, F. Joe Crosswhite, and Jane O. Swafford, "Eighth-Grade Math: An International Study," *Principal,* Vol. 65, No. 1, September, 1985, pp. 37-40.

Travers, Kenneth J., Curtis C. McKnight, and John A. Dossey, "Mathematics Achievement in U.S. High Schools From an International Perspective," *Bulletin,* National Association of Secondary School Principals, November, 1985, pp. 55-63.

123

Appendix V

Table 1

Achievement (Population A)
(Mean percent of items correct)

Country	Arithmetic	Algebra	Geometry	Statistics	Measure-ment
Belgium (Flemish)	58	53	43	58	58
Belgium (French)	57	49	43	52	57
British Columbia	58	48	42	61	52
England/Wales	48	40	45	60	49
Finland	46	44	43	58	51
France	58	55	38	57	60
Hong Kong	55	43	43	56	53
Hungary	57	50	53	60	62
Israel	50	44	36	52	46
Japan	60	60	58	71	69
Luxembourg	45	31	25	37	50
Netherlands	59	51	52	66	62
New Zealand	46	39	45	57	45
Nigeria	41	32	26	37	31
Ontario	55	42	43	57	51
Scotland	50	43	46	59	48
Swaziland	32	25	31	36	35
Sweden	41	32	39	56	49
Thailand	43	38	39	45	48
U.S.A.	51	43	38	57	42

Table 2

Opportunity-to-Learn (Population A)
(Mean percent of items taught)

Country	Arithmetic	Algebra	Geometry	Statistics	Measure-ment
Belgium (Flemish)	76	72	31	39	84
British Columbia	83	84	48	47	77
England and Wales	78	63	54	69	80
Finland	76	70	39	52	70
France	86	87	43	50	92
Hungary	91	91	86	86	97
Israel	71	79	43	52	63
Japan	85	83	51	76	95
Luxembourg	79	52	35	32	82
Netherlands	82	73	67	32	82
New Zealand	67	62	59	60	70
Nigeria	79	72	65	64	71
Ontario	87	70	49	61	84
Sweden	66	45	35	47	67
Thailand	86	83	57	56	86
U.S.A.	87	69	44	73	72

124

Table 3

Achievement (Population B)
(Mean percent of items correct)

Country	Sets and Relations	Number Systems	Algebra	Geometry	Elem. Functions and Calculus	Probability and Statistics
Belgium (Flemish)	72	48	61	42	46	43
Belgium (French)	66	44	55	38	43	42
British Columbia	48	43	47	30	21	38
England/Wales	61	59	66	51	58	64
Finland	77	57	69	48	55	58
Hong Kong	80	78	78	65	71	73
Hungary	35	28	45	30	26	29
Israel	51	46	60	35	45	38
Japan	79	68	78	60	66	70
New Zealand	72	51	57	43	48	58
Ontario	69	47	57	42	46	46
Scotland	50	39	48	42	32	46
Sweden	59	62	60	49	51	64
Thailand	52	33	38	28	26	34
United States	56	40	43	31	29	40

Table 4

Opportunity-to-Learn (Population B)
(Mean percent of items taught)

Country	Sets and Relations	Number Systems	Algebra	Geometry	Elem. Functions and Calculus	Probability and Statistics
Belgium (Flemish)	91	80	92	82	88	46
British Columbia	65	75	82	50	32	29
England/Wales	48	74	86	69	85	87
Finland	88	90	92	79	87	87
Hungary	45	55	87	74	67	27
Israel	38	56	70	49	79	39
Japan	95	80	100	89	92	82
New Zealand	85	90	93	75	93	86
Ontario	62	60	83	52	83	33
Sweden	60	87	90	66	85	75
Thailand	81	75	78	63	63	90
United States	83	83	88	62	54	45

125

Appendix VI
References

Garden, R. A. (1984) *International Sampling Report.* Urbana, Illinois, University of Illinois, mimeo.

Kifer, E. (1986) Issues and implications of differentiated curriculum in the eighth grade. Paper presented at the Annual Meeting of the American Educational Research Association. Lexington, University of Kentucky, mimeo.

Kifer, E. and Robitaille, D. F. (1985) Attitudes, preferences and opinions. In: *International Report on Second International Mathematics Study:* Volume II, to appear.

Miller, M. D. and Linn, R. L. (1985) Cross-national achievement with differential retention rates. Urbana, Illinois, University of Illinois, mimeo.

Quotations

Andy (1986) Sunday Questions. *The Norman* (Oklahoma) *Transcript.*

Broudy, H. S. (1972) *The Real World of the Public Schools,* New York, Harcourt Brace Jovanovich, p. 27.

Bruner, J. S. (1977) *The Process of Education,* Cambridge, Massachusetts, Harvard University Press, p. 13.

Bruner, J. S. (1977a) *op cit,* p. xv.

Carpenter, T., *et al* (1981) Results from the National Assessment of Educational Progress. In: Corbitt, M. K. (ed.) *Results from the Second Mathematics Assessment of the National Assessment of Educational Progress.* Reston, Virginia, National Council of Teachers of Mathematics.

Fey, J. T. and Good, R. A. (1985) Rethinking the sequence and priorities of high school mathematics curricula. In: *The Secondary School Mathematics Curriculum* 1985 Yearbook of the National Council of Teachers of Mathematics, Reston, Virginia, p. 43.

Goodlad, J. I. (1975) *The Dynamics of Educational Change,* New York, McGraw-Hill, p. 95.

Goodlad, J. I. (1984a) *A Place Called School,* New York, McGraw-Hill, p. 166.

Goodlad, J. I. (1984b) *op cit.,* p. 290.

Goodlad, J. I. (1986) Toward a more perfect union, *State Education Leader,* Vol. 5, No. 2, p. 9.

Hill, S. (1985a) Remarks made at the National Conference on Classroom Processes in Secondary School Mathematics, University of Illinois, mimeo.

Hill, S. (1985b) *loc cit.*

Hill, S. (1985c) *loc cit.*

Hill, S. (1985d) *loc cit.*

Quotations (continued)

Lippman, W. (1954) *Atlantic Monthly,* May, p. 38.

Mathematical Sciences Education Board (1986a) Report prepared for fall meeting, mimeo.

Mathematical Sciences Education Board (1986b) *op cit.*

McMurrin, S. (1963a) The curriculum and the purposes of education. In: Heath, R. W. (ed.) *New Curricula,* Harper and Row, New York, p. 266.

McMurrin, S. (1963b) *loc cit.*

National Science Board (1983a) *Educating Americans for the 21st Century,* National Science Foundation, Washington, D.C., p. v.

National Commission on Excellence in Education (1983a) *A Nation at Risk,* U.S. Government Printing Office, Washington, D.C., p. 7.

National Commission on Excellence in Education (1983b) *op cit.,* p. 13.

National Commission on Excellence in Education (1983c) *op cit.,* p. 7.

National Commission on Excellence in Education (1983d) *loc cit.*

National Council of Teachers of Mathematics (1986) Project proposal, mimeo.

Phenix, P. H. (1958) *Philosophy of Education,* Holt, Rinehart and Winston, New York, p. 59.

Ravitch, D. (1985) The problem of educational reform. In: Ravitch, D., *The Schools We Deserve,* Basic Books, New York, p. 16.

Resnick, D. P. and Resnick, L. B. (1985) Standards, curriculum and performance: A historical and comparative perspective, *Educational Researcher,* April, p. 10.

Steen, L. A. (1984) Remarks made at the National Conference on the U.S. Findings of the Second International Mathematics Study, University of Illinois.

Usiskin, Z. (1985) We need another revolution in secondary school mathematics. In: *The Secondary School Mathematics Curriculum,* 1985 Yearbook of the National Council of Teachers of Mathematics, Reston, Virginia, p. 18.

Whitehead, A. N. (1916) Presidential address to the Mathematical Association of England, *The Aims of Education,* Mentor Books, Copyright 1929, New York, p. 26.